Become a Python Data Analyst

Perform exploratory data analysis and gain insight into scientific computing using Python

Alvaro Fuentes

BIRMINGHAM - MUMBAI

Become a Python Data Analyst

Commissioning Editor: Pravin Dhandre
Acquisition Editor: Namrata Patil
Content Development Editor: Athikho Sapuni Rishana
Technical Editor: Kushal Shingote
Copy Editor: Safis Editing
Project Coordinator: Kirti Pisat
Proofreader: Safis Editing
Indexer: Priyanka Dhadke
Graphics: Jisha Chirayil
Production Coordinator: Arvindkumar Gupta

First published: August 2018

Production reference: 1310818

Published by Packt Publishing Ltd.
Livery Place
35 Livery Street
Birmingham
B3 2PB, UK.

ISBN 978-1-78953-170-1

www.packtpub.com

`mapt.io`

Mapt is an online digital library that gives you full access to over 5,000 books and videos, as well as industry leading tools to help you plan your personal development and advance your career. For more information, please visit our website.

Why subscribe?

- Spend less time learning and more time coding with practical eBooks and Videos from over 4,000 industry professionals

- Improve your learning with Skill Plans built especially for you

- Get a free eBook or video every month

- Mapt is fully searchable

- Copy and paste, print, and bookmark content

Packt.com

Did you know that Packt offers eBook versions of every book published, with PDF and ePub files available? You can upgrade to the eBook version at `www.packt.com` and as a print book customer, you are entitled to a discount on the eBook copy. Get in touch with us at `customercare@packtpub.com` for more details.

At `www.packt.com`, you can also read a collection of free technical articles, sign up for a range of free newsletters, and receive exclusive discounts and offers on Packt books and eBooks.

Contributor

About the author

Alvaro Fuentes is a data scientist with an M.S. in quantitative economics and applied mathematics with more than 10 years of experience in analytical roles. He worked in the central bank of Guatemala as an economic analyst, building models for economic and financial data. He founded Quant to provide consulting and training services in data science topics and has been a consultant for many projects in fields such as business, education, psychology, and mass media. He has taught courses to students in topics such as data science, mathematics, statistics, R programming, and Python. He also has technical skills in R programming, Spark, PostgreSQL, Microsoft Excel, machine learning, statistical analysis, econometrics, and mathematical modeling.

Packt is searching for authors like you

If you're interested in becoming an author for Packt, please visit authors.packtpub.com and apply today. We have worked with thousands of developers and tech professionals, just like you, to help them share their insight with the global tech community. You can make a general application, apply for a specific hot topic that we are recruiting an author for, or submit your own idea.

Table of Contents

Preface

Python is one of the most common and popular languages used by leading data analysts and statisticians for working with massive datasets and complex data visualizations.

Become a Python Data Analyst introduces Python's most essential tools and libraries that you need to work with the data analysis process, right from preparing data to performing simple statistical analyses and creating meaningful data visualizations.

In this book, we will cover Python libraries such as NumPy, pandas, matplotlib, seaborn, SciPy, and scikit-learn, and apply them in practical data analysis and statistics examples. As you make your way through the chapters, you will learn to efficiently use the Jupyter Notebook to operate and manipulate data using the NumPy and pandas libraries. In the concluding chapters, you will gain experience in building simple predictive models, statistical computation and analysis using rich Python tools, and proven data analysis techniques.

By the end of this book, you will have hands-on experience of performing data analysis with Python.

Who this book is for

Become a Python Data Analyst is for entry-level data analysts, data engineers, and BI professionals who want to make complete use of Python's tools for performing efficient data analysis. Prior knowledge of Python programming is necessary to understand the concepts covered in this book.

What this book covers

Chapter 1, *The Anaconda Distribution and Jupyter Notebook*, covers the most important libraries for data science with Python. This is a well-charted overview of the main objects, attributes, methods, and functions that we will use for doing predictive analytics with Python.

Chapter 2, *Vectorizing Operations with NumPy*, explores Numpy—this is the library upon which almost all other scientific computing in Python projects are based. Learning how to handle NumPy arrays is crucial for doing anything related to data science in Python.

Chapter 3, *Pandas - Everyone's Favorite Data Analysis Library*, gives an overview of pandas which is a library that provides high performance, easy-to-use data structures, and data analysis tools for the Python programming language. We data scientists love it, and it is one of the key reasons behind Python's popularity in the data science community. In this section, we show by example how to perform descriptive analysis with pandas.

Chapter 4, *Visualization and Explanatory Data Analysis*, explains that visualization is a key topic for data science. Python provides a lot of options for doing visualizations for different purposes. In this volume, we learn about two of the most popular libraries, matplotlib and seaborn, and perform exploratory data analysis on real-world datasets.

Chapter 5, *Statistical Computing with Python*, explains how to perform common statistical computations with Python and use them to make sense of a dataset that contains information about the alcohol consumption of teenagers.

Chapter 6, *Introduction to Predictive Analytics Models*, gives a brief introduction to predictive analytics and builds a model to predict the drinking habits of teenagers.

To get the most out of this book

This book will give you an overview of data analysis in Python. This will take you through the main libraries of Python's data science stack. It will explain how to use various Python tools to analyze, visualize, and process data effectively.

Download the example code files

You can download the example code files for this book from your account at www.packt.com. If you purchased this book elsewhere, you can visit www.packt.com/support and register to have the files emailed directly to you.

You can download the code files by following these steps:

1. Log in or register at www.packt.com.
2. Select the **SUPPORT** tab.
3. Click on **Code Downloads & Errata**.
4. Enter the name of the book in the **Search** box and follow the onscreen instructions.

Once the file is downloaded, please make sure that you unzip or extract the folder using the latest version of:

- WinRAR/7-Zip for Windows
- Zipeg/iZip/UnRarX for Mac
- 7- Zip/PeaZip for Linux

The code bundle for the book is also hosted on GitHub at `https://github.com/PacktPublishing/Become-a-Python-Data-Analyst`. In case there's an update to the code, it will be updated on the existing GitHub repository.

We also have other code bundles from our rich catalog of books and videos available at `https://github.com/PacktPublishing/`. Check them out!

Download the color images

We also provide a PDF file that has color images of the screenshots/diagrams used in this book. You can download it here: `http://www.packtpub.com/sites/default/files/downloads/BecomeaPythonDataAnalyst_ColorImages.pdf`.

Conventions used

There are a number of text conventions used throughout this book.

`CodeInText`: Indicates code words in text, database table names, folder names, filenames, file extensions, pathnames, dummy URLs, user input, and Twitter handles. Here is an example: "Mount the downloaded `WebStorm-10*.dmg` disk image file as another disk in your system."

A block of code is set as follows:

```
# The largest heading
## The second largest heading
###### The smallest heading
```

When we wish to draw your attention to a particular part of a code block, the relevant lines or items are set in bold:

```
[default]
exten => s,1,Dial(Zap/1|30)
exten => s,2,Voicemail

(u100)
exten => s,102,Voicemail(b100)
exten =>

i,1,Voicemail(s0)
```

Bold: Indicates a new term, an important word, or words that you see onscreen. For example, words in menus or dialog boxes appear in the text like this. Here is an example: "Click on **Next** in the first installer dialog box..."

 Warnings or important notes appear like this.

 Tips and tricks appear like this.

Get in touch

Feedback from our readers is always welcome.

General feedback: If you have questions about any aspect of this book, mention the book title in the subject of your message and email us at `customercare@packtpub.com`.

Errata: Although we have taken every care to ensure the accuracy of our content, mistakes do happen. If you have found a mistake in this book, we would be grateful if you would report this to us. Please visit `www.packt.com/submit-errata`, selecting your book, clicking on the Errata Submission Form link, and entering the details.

Piracy: If you come across any illegal copies of our works in any form on the Internet, we would be grateful if you would provide us with the location address or website name. Please contact us at `copyright@packt.com` with a link to the material.

If you are interested in becoming an author: If there is a topic that you have expertise in and you are interested in either writing or contributing to a book, please visit `authors.packtpub.com`.

Reviews

Please leave a review. Once you have read and used this book, why not leave a review on the site that you purchased it from? Potential readers can then see and use your unbiased opinion to make purchase decisions, we at Packt can understand what you think about our products, and our authors can see your feedback on their book. Thank you!

For more information about Packt, please visit `packt.com`.

The Anaconda Distribution and Jupyter Notebook

1

In this book, you will learn the basic concepts of data analysis using Python. In the first chapter, we will learn how to install the Anaconda distribution, which contains all the software needed for this book. We will also get to know Jupyter Notebook, which is the computing environment where we will do all of our work. Something nice about this book is that we take a hands-on practical approach that will help you to master the tools very effectively.

The following are the main topics that will be covered in this chapter:

- The Anaconda distribution and the problems it solve
- How to install it in our computer and get ready to start working
- Jupyter Notebook, where we will perform our computations and analysis
- Some useful commands and keyboard shortcuts used in Jupyter Notebook

The Anaconda distribution

Anaconda is a free, easy-to-install package management and environment management Python distribution created for developers and data scientists to make package management and deployment in scientific computing, data science, statistical analysis, and machine learning an easy task. It is software that is produced and distributed by Continuum Analytics and is available for free download at `https://www.anaconda.com/download/`.

Anaconda is a toolbox, a ready-to-use collection of related tools for doing data analytics with Python. The individual tools are also available for free download, but it is definitely more convenient to get the whole toolbox. This is the main problem that Anaconda solves, saving you the time it takes to look for every individual tool and install them on your system. In addition, Anaconda also takes care of package dependencies and other potential conflicts and problems that are the outcome of installing Python packages individually.

Installing Anaconda

On opening the previously mentioned URL, we are given download options for each operating system. So, here, you must look for the appropriate installer for your operating system. You will see two installers: one is for Python 3.6 and the other is for Python 2.7. In this book, we will be using Python 3.6:

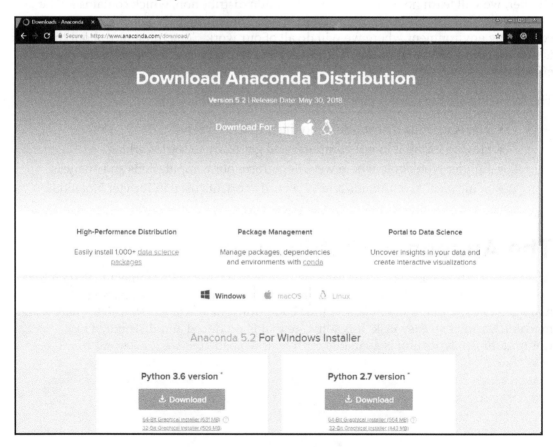

Let's download the latest version of the Anaconda software and save in our `Downloads` folder.

 Here, we choose the 64-bit installer of Anaconda 5.2 for Windows, since we will be working on a Windows environment in this book. Installation for macOS and Linux will also be similar.

Installing Anaconda is very simple; it is no different from any other software you have installed before. Double-click on the `.exe` file and install the software on your system. The steps are easy and you'll get the normal prompts you see when installing software:

1. Click on **Next** in the first installer dialog box.
2. Then, we have the license agreement, where we can click on **I Agree** after going through the terms and conditions of the software.
3. From the options, select **Just Me** and click **Next**.
4. Select the default installation destination folder and click **Next**.
5. Next, it asks you for the environment variables and whether you want to register Anaconda as your default Python. Check both boxes and click on **Install**.
6. Once the installation is complete, click on **Finish** in the installer dialog box.

Jupyter Notebook

Jupyter Notebook is a web application that allows you to create and share documents that contain live code, equations, visualizations, and explanatory text. It is the application we will be working with in this book. Its uses include data cleaning and transformation, numerical simulation, statistical modeling, machine learning, and much more. Jupyter Notebook is similar to a canvas, or an environment, that allows you to use a programming language, in our case Python, to perform computations and to show the results in a very convenient way.

It is very convenient if you're doing some kind of analytical work, because often you want to include explanatory text, the code that produced the results, and the visualizations, which are shown in Jupyter Notebook. So it is a very convenient way of doing analytical work with any programming language, especially with Python. The Jupyter project was born out of the IPython project in 2014. Now, it has evolved to support interactive data science and scientific computing across many other programming languages, so you can use the Jupyter Notebook with many other programming languages (up to almost 20 languages). The name Jupyter comes from Julia, Python, and R, which were the three programming languages that were first supported.

Creating your own Jupyter Notebook

To start Anaconda and open a Jupyter Notebook, we click **Anaconda Prompt** from our list of installed programs. Anaconda Prompt is a Terminal where you can type commands. But first, let's create a folder called `PythonDataScience` on our desktop. This will be the directory where all the Python code that we code and run for this book in Jupyter Notebook will be stored.

Once the Terminal is open, navigate to `PythonDataScience` by typing the command `cd Desktop/PythonDataScience` and pressing *Enter*. To start the Jupyter Notebook application inside this directory, type the command `jupyter notebook` and hit *Enter*. This will start the application and you will see the main screen of the application opened in a tab on your browser:

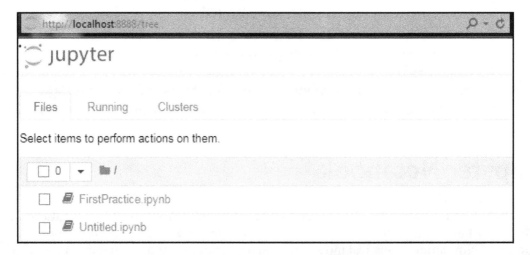

You have three tabs here. One is the **Files** tab, where you will see all the files that you have inside the folder where you started the application. On the **Running** tab, you will see the things that you have running, either Terminals or Notebooks. The **Clusters** tab displays details about parallel computing, but we will not be using this feature in this book.

The main tab that we will be using in this book is the **Files** tab. To create a new Jupyter Notebook, go to **New | Python 3 Notebook**:

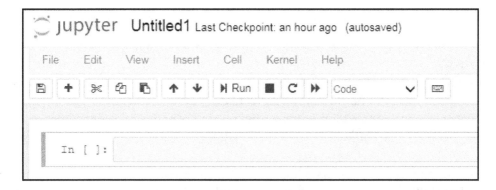

This will start a new file, which is the Jupyter Notebook where you can start coding and running your Python code.

Notebook user interfaces

Jupyter Notebook has some very useful user interfaces that display important information and notifications while you work on the Notebook. Let's go to **Help** and click on the first option, **User Interface Tour**, to take a quick look and get familiar with the interfaces of Jupyter Notebook:

The following are the main interfaces you will find on the main page of the Notebook:

- **Title (1)**: This is your filename and you can also change the filename of your Notebook.
- **Menu bar (2):** Similar to any other desktop application, you have a menu bar where you can find different actions to do with the Notebook.
- **Toolbar (3):** This is located below the menu bar and contains little icons to perform some commonly performed actions, such as saving a file, cutting cells, pasting cells, moving cells, and so on.
- **Mode indicator (4):** This is located on the right-hand side of the menu bar. The Notebook has two modes, Edit mode and Command mode. Command mode has many keyboard shortcuts that you can use. In this mode, no icon is displayed in the indicator area, and the actions you can perform in this mode are actions that have to do with the file itself, such as saving the file, copying and pasting cells, and so on. Edit mode allows you to write code or text in a cell, if you are in a cell. When you are in Edit mode, you will see a little pencil in the indicator area.

 A Jupyter Notebook is composed of two types of cells, code cells and text cells. The border of a selected cell will be green when you are in Edit mode. To go back to Command mode from Edit mode, you can press the *Esc* key or *Ctrl + M*. There are many keyboard shortcuts that you can use and there is a list in the **Help** menu.

- **Kernel indicator (5):** This displays the status of the system's computation progress. To interrupt the computations that are in progress, you can use the stop button located on the toolbar.
- **Notification area (6):** This area will display messages such as **saving the file**, or **interrupting the kernel**, and so on, and you will see the actions that you're performing in the notification area.

Using the Jupyter Notebook

Lets open a new Jupyter Notebook and create a new Python 3 Notebook, then name the Notebook as `FirstPractice`. As mentioned earlier, the Notebook is made out of cells, and you have two types of cells: the default cell type, which is called the **code** cell, and the other type of cells are **text** cells. We have a code cell every time we open a Notebook where you can execute any Python statement.

Running code in a code cell

We will run a few simple code statements and learn how to run those statements, and also learn how to change a cell type from code to text and vice versa. Let's execute the first code by typing 1 + 1 in the first code cell, and if you run the code of the cell with the run cell button, you will see the following output in the line below the code cell:

Next, let's create a variable, a, assign its value as 10, and run the code. Now this variable has been created, but since we didn't create any code to compute the variable, we won't see any output. But the statement was run, and now, if you use this variable, add 1 to it, and run the code, you will see the following result:

```
In [9]:  a=10

In [10]: a+1

Out[10]: 11
```

Now, let's see an example of the for syntax using a variable, i:

```
In [13]: for i in range(10):
             print(i)

         0
         1
         2
         3
         4
         5
         6
         7
         8
         9
```

The code will tell the Notebook to print the value of `i` if and when the value is within the range of 10, which gives the preceding result.

Running markdown syntax in a text cell

As mentioned, the default type of each cell is a code cell, where we write Python expressions. The other type of cell that we have is a text cell, and a text cell is used to actually write text. In the cell below the output, let's type `This is regular text`. To tell the Notebook that this is not Python code and this is actually some text, you go to **Cell | Cell Type | Markdown**. Let's run this now and you will find that what you get as output is just the text, the same text we entered:

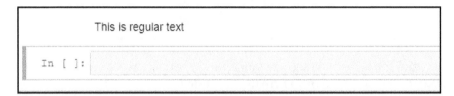

Jupyter also allows us to format the text in many ways by using markdown syntax. If you are not familiar with markdown, you can go to **Help | Markdown**, where you will be taken to one of the GitHub help pages.

 The markdown that you can use in Jupyter is the same markdown you use in GitHub.

There are many ways you can style and format your text; you can find all the information at: `https://help.github.com/articles/basic-writing-and-formatting-syntax/`. For this chapter, we will just look at the headings that are very important.

To create a heading, we add one to six # symbols before the heading text. The number of # symbols determines the size of the heading, starting from one to six # symbols, the largest to smallest heading, as follows:

```
# The largest heading
## The second largest heading
###### The smallest heading
```

The following screenshot shows the output for the preceding syntax:

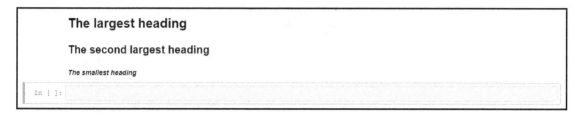

If you run them in a code cell, you will get a bunch of Python commands, but we know that we want to see these as formatted text, so we need to tell the Notebook that these are actually texts by marking the cell type as markdown, and when you run the cell you get the preceding result.

Styles and formats

You can introduce other formats, such as bold, italic, strikethrough, and bold and italic. The following table shows the different styles and their corresponding syntax with an example of each:

Style	Syntax	Keyboard shortcut	Example	Output
Bold	`** **` or `__ __`	command/control + b	`**This is bold text**`	**This is bold text**
Italic	`* *` or `_ _`	command/control + i	`*This text is italicized*`	*This text is italicized*
Strikethrough	`~~ ~~`		`~~This was mistaken text~~`	~~This was mistaken text~~
Bold and italic	`** **` and `__ __`		`**This text is _extremely_ important**`	**This text is _extremely_ important**

You can also introduce quotes, for which the syntax is denoted by, the > symbol. Run a markdown cell with the following syntax:

```
In the words of Abraham Lincoln:
> Pardon my French
```

This will give the following result:

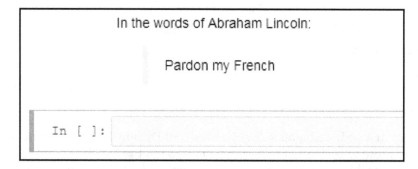

We find the style and format of the texts remains the same, similar to the line of text preceding the >, but the quoted text is a little indented from the normal indentation.

Lists

You can use – or * before one or more lines to create a bulleted list. You can also use 1, 2, 3, and so on, to create a numbered or ordered list:

```
-  George Washington
-  John Adams
-  Thomas Jefferson

1.  James Madison
2.  James Monroe
3.  John Quincy Adams
```

On running the markdown syntax, we get the following result:

- George Washington
- John Adams
- Thomas Jefferson

1. James Madison
2. James Monroe
3. John Quincy Adams

You will get a lot more styling and formatting syntax from a website that you can use in the Notebook.

Useful keyboard shortcuts

It is very annoying to use the mouse every time you want to run the cell or to transform a code cell into a markdown cell. To ease these tasks, we have a lot of keyboard shortcuts that you can use in Jupyter. Let's look at some of the most important ones.

To run a cell, for example, if you want to run the computation of 1 + 2, you can hit *Alt + Enter*. This shortcut will run the code in the cell and create a new cell below the output line. But if you just want to run the computation, you will press *Ctrl + Enter*; this will get the Notebook to run the code in that cell and show you the output, but no new cell will be created. If you want to insert a new cell below the current cell, you can press *B*, and then pressing *A* will create new cells above the current cell. This is very useful when you're working interactively within the Notebook and you want to create many cells to have more room to work.

 If you find yourself using a command very often and you want to learn the shortcut key; you can go to **Help|Keyboard Shortcuts**. You will see a list of all the keyboard shortcuts that you can use in the Notebook.

Another very useful thing is how to transform cells from code cells to markdown cells. If you want to transform a code cell into a text cell, press the *M* key; this will get you from Edit mode to Command mode. Also, if you are in Command mode and you want to transform the cell into a code cell or Edit mode, press the *Esc* key.

Summary

In this chapter, we learned about the Anaconda distribution. We also installed all the software that we will use in the course, which is contained in the Anaconda distribution, and we learned about Jupyter Notebook and the basics of working with Jupyter Notebooks; about the code cells, the markdown cells, and the most useful keyboard command shortcuts to work with Jupyter and make our workflow more efficient.

In the next chapter, we will learn about NumPy, which is the core library for doing numerical computing and is at the core of every scientific computing project in the PythonX system.

Vectorizing Operations with NumPy 2

In this chapter, we will learn about **Numerical Python** (**NumPy**) extensions, which is a library for the Python programming language, what it is, and why we need it. We will also look at arrays, which are the most important type of objects in the `numpy` library. We will learn how to work with arrays, what the most important methods are, and the attributes that we can use with arrays. Then, we will apply our knowledge and do some simulations to see how we use **NumPy** in the real world. By the end of this chapter, you will know all the foundations that you need to work with other libraries in Python's Data Science Stack, such as **Matplotlib**. We will also get into some motivating examples to see why we need NumPy and the main problem it solves.

We will cover the following topics:

- Introduction to NumPy
- NumPy arrays – creation, methods, and attributes
- Basic math with arrays
- Manipulation with arrays
- Using NumPy for simulations

Introduction to NumPy

NumPy, also known as Python's vectorization solution, is the fundamental package for doing scientific computing with Python. It gives us the ability to create multidimensional array objects and to perform faster mathematical operations than we can do with base Python. It is the basis of most of Python's Data Science ecosystem. Most of the other libraries that we use in data analytics with Python, such as `scikit-learn` and `pandas` rely on NumPy. Some advanced features of NumPy are as follows:

- It provides sophisticated (broadcasting) functions
- It provides tools for integrating with lower-level languages such as C, C++, and Fortran
- It has the ability to do linear algebra and complex mathematical operations such as **Fourier Transform (FT)** and **random number generator (RNG)**

So, if you need to do some really high-performance data analysis at scale and you need the code to run very fast, you can integrate Python code with these lower-level languages.

Problems and solutions

Let's now look into the type of problem that NumPy solves and why we need to use it, with a small example. Let's say that we have some data about distances and times, and we want to do some processing with the data to calculate the speeds. To do this, one approach might be to create an empty list in Python and then to write a `for` loop, and the `for` loop will append the operation, the distance over the time and give its speeds, as shown in the following code:

```
In [1]: distances = [10, 15, 17, 26, 20]
        times = [0.3, 0.47, 0.55, 1.20, 1.0]

In [2]: # Calculate speeds with Python
        speeds = []
        for i in range(len(distances)):
            speeds.append(distances[i]/times[i])

        speeds

Out[2]: [33.333333333333336,
         31.914893617021278,
         30.909090909090907,
         21.666666666666668,
         20.0]
```

The preceding screenshot shows the new list that was created by doing the operation, the distance over time. There is a better way to do it, a more Pythonic way, which is to do a list comprehension. So if we use `[d/t for d,t in zip(distances, times)]` instead of the `for` loop shown in the following screenshot, you will get the same result:

```
In [3]:  [d/t for d,t in zip(distances, times)]

Out[3]:  [33.333333333333336,
          31.914893617021278,
          30.909090909090907,
          21.666666666666668,
          20.0]
```

This is the traditional way we would solve the problem to calculate speeds for the given data, and the earlier way is the approach that we will take in base Python.

Let's see another example via an analysis of shopping: given the product quantities and the prices for each product, we have to get the total sum of the purchase made. So, to get the total of the purchase, we need to multiply every quantity by the respective price, and after doing all the multiplications, we have to perform an addition. The following is the code to do it in regular Python:

```
sum([q*p for q,p in zip(product_quantities, prices)])
```

First, we apply the `sum` function, which is quantity times the price, and then, to get the addition, we generate a list comprehension that we usually use when we have to solve these kinds of problem in base Python. So if we run the cell, we get the result and the total is `157.1`. This is the total cost of the set of quantities and prices:

```
In [4]:  product_quantities = [13, 5, 6, 10, 11]
         prices = [1.2, 6.5, 1.0, 4.8, 5.0]
         total = sum([q*p for q,p in zip(product_quantities, prices)])
         total

Out[4]:  157.1
```

It would be really nice if we could have the ability to just get the speeds by `distances/times` and define the total in the second problem as just `product_quantities` times the `prices` list, and get the sum of the multiplication.

But if we run them on the cell, we will get an error because base Python doesn't allow division of a list by a list and multiplication of a list by a list. So this is precisely the kind of operation and the kind of mathematical operation that we can do with NumPy, and this is what we mean when we say vectorization. Vectorization means operating an array, such as object or list, or doing operations element by element.

NumPy arrays

NumPy's main object is a homogeneous multidimensional array. An array is essentially a table of elements (usually numbers), all of the same type, indexed by a tuple of positive integers. The index in NumPy arrays is zero-based, so the first element is the 0th element; the second element is the 1st element, and so on. In NumPy, dimensions are called **axes** and the number of axes, or dimensions, is called the **rank** or **dimension** of the array. To import NumPy into our Jupyter Notebook, we use the `numpy as np` convention import.

Creating arrays in NumPy

There are the following two methods to create arrays in Python:

- Creating arrays from lists
- Using the built-in functions that NumPy provides

Creating arrays from lists

To create a NumPy array from a list, we use the `np.array` function. The lists that we saw in our earlier examples, such as `distances`, `times`, `product_quantities`, and `prices` will be transformed into arrays using the `np.array` function. To do so, we can run the following lines of code for each list:

```
distances = np.array(distances)
times = np.array(times)
product_quantities = np.array(product_quantities)
prices = np.array(prices)
```

This will transform this object from Python lists into NumPy arrays. Now if you take a look at this new object by calling the `distances` object name in the cell and running it, you will see that it is, in fact, an array as shown in the following screenshot:

```
In [9]:  distances

Out[9]:  array([10, 15, 17, 26, 20])
```

We can also ask Python for the type of the object by using the `type(distances)` code. On running this code in a cell, we will get the following output that displays the object type:

```
In [10]:  type(distances)

Out[10]:  numpy.ndarray
```

Python shows us an object of the `numpy.ndarray` type; here nd stands for an n-dimensional array, also called a one-dimensional array, and sometimes we refer to them as vectors.

 A NumPy vector and a one-dimensional array are the same. If we pass `np.array()` (a list of lists), it will create a two-dimensional array. If we pass a list of lists, it will create a three-dimensional array, and so on.

Let's take a look at another example to have a better understanding of how to create a two-dimensional array by passing a list of lists:

```
In [11]:  A = np.array([[1, 2], [3, 4]])
          A

Out[11]:  array([[1, 2],
                 [3, 4]])
```

Here, we have a list that contains two elements and each element is a list itself within the outer list, and every list has two elements. So this will be a two-dimensional array with two rows and two columns, and sometimes we refer to two-dimensional arrays as matrices. The preceding example is a matrix of size 2 x 2.

Creating arrays from built-in NumPy functions

NumPy also provides us with some functions to create arrays, because sometimes we need to initialize arrays even though we don't have the values of the elements of the array. So, it is very useful to create arrays with default values. Let's look into each of the important functions.

First, let's see an example of the np.zeros function, which will create a NumPy array filled with zeros. You can optionally specify the type of the array, so for this instance, we will specify the argument type as dtype=int to specify that all the elements of this array will be of the integer type as shown in the following screenshot:

```
In [12]:  # Create a length-10 integer array filled with zeros
          np.zeros(10, dtype=int)

Out[12]:  array([0, 0, 0, 0, 0, 0, 0, 0, 0, 0])
```

For the preceding example, we have np.zeroes(10, dtype=int) to create an array length of 10 integers and, as the result, we have a NumPy array of the integer type with 10 zeros.

Next, we have the np.ones function, which gives us the option to pass the shape argument, which will be the number of elements that we will have in each dimension. We will assign the first dimension value as 3 and the second dimension value as 5. The arguments type will be float:

```
In [13]:  # Create a 3x5 floating-point array filled with ones
          np.ones(shape=(3, 5), dtype=float)

Out[13]:  array([[1., 1., 1., 1., 1.],
                 [1., 1., 1., 1., 1.],
                 [1., 1., 1., 1., 1.]])
```

On running np.ones(3, 5), dtype=float), we get a 3 x 5 matrix filled with 1 for each object of the float type or floating-point.

There is another very useful function in NumPy where you can create arrays filled with a linear sequence. There are the following two ways to create linear sequences:

- Using the np.arange function, you can provide a starting point, a stopping point or number, and a step value for each iteration. If you don't provide a step, the default value will be 1. Take a look at the following screenshot:

```
In [14]:  # Create an array filled with a linear sequence
          # Starting at 0, ending at 20, stepping by 2
          np.arange(start=0, stop=20, step=2)

Out[14]:  array([ 0,  2,  4,  6,  8, 10, 12, 14, 16, 18])
```

For the preceding example, we called a sequence whose start point is 0 and whose stop point is 20 with a step value of 2. On running the cell, we get a NumPy array that will start with 0 and it will go in sizes of step value 2. This will go on until it gets to the stopping point. The stopping point, or stopping value, won't be included here.

- Another way to create a liner sequence is by using the np.linspace function. For this, we provide the lower value, the upper value, and the number of values to be evenly spaced between the lower and the upper value. In this function, the upper value will be included, unlike the np.arange function as shown in the following screenshot:

```
In [15]:  # Create an array of 20 values evenly spaced between 0 and 1
          np.linspace(0, 1, 20)

Out[15]:  array([0.        , 0.05263158, 0.10526316, 0.15789474, 0.21052632,
                 0.26315789, 0.31578947, 0.36842105, 0.42105263, 0.47368421,
                 0.52631579, 0.57894737, 0.63157895, 0.68421053, 0.73684211,
                 0.78947368, 0.84210526, 0.89473684, 0.94736842, 1.        ])
```

For this example, we get a sequence that starts from 0 up to 1, including 1; all the values in between are evenly spaced, and we have 20 values in total.

Attributes of arrays

There are many attributes of arrays in NumPy; we will look into three important and commonly used attributes in Python. To see the three attributes, let's first create an array of *3 x 4* of the `float` type, and all the values as 1 as shown in the following screenshot:

```
In [16]:  A = np.ones(shape=(3, 4), dtype=float)
          A

Out[16]:  array([[1., 1., 1., 1.],
                 [1., 1., 1., 1.],
                 [1., 1., 1., 1.]])
```

Let's take a look at the following three attributes. The first attribute that we will see is the number of dimensions. We use the `A.ndim` attribute to check the dimension of the array, and for this example, we have the number of dimensions in the array as 2. To know the shape, we use the `A.shape` attribute, which will give us the number of values in every dimension; the first dimension has 3 values and the second dimension has 4 values. And finally, the size is the total number of elements that we have in the array. We use the `A.size` attribute, which gives us 12, the total number of elements in the array:

Basic math with arrays

Let's do some basic math with our arrays using the values for `distances` and `times` from the lists that we used earlier to introduce the NumPy array. We said that it would be wonderful if we had the opportunity to calculate `speeds`, just by dividing `distances` over `times`. These are the kind of vectorized operations that can be done with NumPy arrays. If we define the `speeds` object as `distances/times`, NumPy will perform this division operation element by element, which is why it is called a vectorized operation:

```
In [17]:  A.ndim

Out[17]:  2

In [18]:  A.shape

Out[18]:  (3, 4)

In [19]:  A.size

Out[19]:  12
```

On running the cell, NumPy will calculate the values of 10/0.3, 15/0.47, and so on, and we will get the vector of speeds in a NumPy array. The other example that we wanted to operate with was to get the sum of `product_quantities` and `prices`. To do so, let's create another NumPy array called `values`, which will be the sum of `product_quantities` times `prices`, and then to get the total, we add up all the elements in the `values` vector as shown in the following screenshot:

```
In [20]:  speeds = distances/times
          speeds

Out[20]:  array([33.33333333, 31.91489362, 30.90909091, 21.66666667, 20.      ])
```

On running the cell, we get the values of 15.6 (which is the value of 13*1.2), 32.5 (which is the value of 5*6.5), and so on, which is the vector of `values`; and to get the total, which is the sum of all the elements in `values`, we just use the `sum` attribute. These are the kind of operations that we can do with NumPy.

Let's see one more example, by creating another vector or NumPy array: x with the `start` value as 0, `stop` value as 20, and the `step` value as 2:

```
In [21]:  values = product_quantities*prices
          total = values.sum()
          print(values)
          total

          [15.6 32.5  6.  48.  55. ]

Out[21]:  157.1
```

On running the cell, we get the numbers from 0 to 18, and when you do basic operations with the NumPy array, usually the operations will be performed element by element. Take a look at the following screenshot:

```
In [22]:  x = np.arange(start=0, stop=20, step=2)
          x

Out[22]:  array([ 0,  2,  4,  6,  8, 10, 12, 14, 16, 18])
```

When you add 1 to these vectors, you see that the result is adding 1 to each element of the original array. Similarly, when you multiply by 2, you will see that the result is two times every element of the array. Likewise, you can perform all the essential and basic arithmetic operations such as addition, subtraction, multiplication, and division.

Another very useful thing about NumPy is that it provides us with what is called **universal functions**. Universal functions are mathematical functions that we can use in arrays, and when we use these functions in arrays, it will be applied element by element. For instance, if we want to know `sin` of each of the elements of x, or if we want to know the exponent of each of the element of x, we can do so by running `np.sin(x)` and `np.exp(x)`. On running the cells, we get the following output, where the `sin` function is applied in every element. Similarly, we have the exponential function applied to all the elements of x:

```
In [23]:  x + 1

Out[23]:  array([ 1,  3,  5,  7,  9, 11, 13, 15, 17, 19])

In [24]:  x * 2

Out[24]:  array([ 0,  4,  8, 12, 16, 20, 24, 28, 32, 36])

In [25]:  x/2

Out[25]:  array([0., 1., 2., 3., 4., 5., 6., 7., 8., 9.])
```

Likewise, we can apply the logarithm and combine operations. In the following screenshot, we have the natural logarithm for the values of x+1 and the square root of every element in the x array:

```
In [26]:  # Universal functions
          np.sin(x)

Out[26]:  array([ 0.        ,  0.90929743, -0.7568025 , -0.2794155 ,  0.98935825,
                 -0.54402111, -0.53657292,  0.99060736, -0.28790332, -0.75098725])

In [27]:  np.exp(x)

Out[27]:  array([1.00000000e+00, 7.38905610e+00, 5.45981500e+01, 4.03428793e+02,
                 2.98095799e+03, 2.20264658e+04, 1.62754791e+05, 1.20260428e+06,
                 8.88611052e+06, 6.56599691e+07])
```

Common manipulations with arrays

Now let's talk about some common manipulations that you can do with NumPy arrays, and we will talk about three of them such as indexing, slicing, and reshaping.

Indexing arrays

Indexing is just getting and setting the value of individual array elements. If you want to access individual elements in the array, you do it in the same way you access elements in a Python list. To see how each indexing call is obtained, let's first create a one-dimensional array using `np.linspace`. If you want to access the first element of the array, you can use the zero-based index as shown in the following screenshot:

```
In [28]:  np.log(x+1)

Out[28]:  array([0.       , 1.09861229, 1.60943791, 1.94591015, 2.19722458,
          2.39789527, 2.56494936, 2.7080502 , 2.83321334, 2.94443898])

In [29]:  np.sqrt(x)

Out[29]:  array([0.       , 1.41421356, 2.       , 2.44948974, 2.82842712,
          3.16227766, 3.46410162, 3.74165739, 4.       , 4.24264069])
```

For the zero-based index, we run the cell with `one_dim[0]` ; this will give us the element with index 0. So in the preceding array, the first element is -0.5 ; we can also find an element at each index by replacing 0 with the element index we want. If you want to change the value of an element, you can do so by getting the element using the index and then assign the value. For example, let's say that we want to change the first element of the vector with index 0 to the value of 1; all you need to do is get the element by using the index and then assign the new value. The following screenshot depicts how the first element, -0.5, is changed to 1:

```
In [59]:  one_dim = np.linspace(-0.5, 0.6, 12)
          one_dim

Out[59]:  array([-0.5, -0.4, -0.3, -0.2, -0.1,  0. ,  0.1,  0.2,  0.3,  0.4,  0.5,
          0.6])

In [60]:  one_dim[0]

Out[60]:  -0.5
```

Now when you are working with two-dimensional arrays, the indexing operation has to be done with two indices, because you have one index for every dimension. First, we create a two-dimensional array. Then, if you want to get the value of the fourth element from the row with index 0, which is the first element in the column with index 3, we use the `two_dim[0,3]` code; the first dimension 0 is the index of the row and for the second dimension 3 is the index for the column:

```
In [32]:  one_dim[0] = 1
          one_dim

Out[32]:  array([ 1. , -0.4, -0.3, -0.2, -0.1,  0. ,  0.1,  0.2,  0.3,  0.4,  0.5,
                  0.6])
```

We receive the preceding output and, just as we did with one-dimensional arrays, you can change the value of any element in a two-dimensional array. In this case, we are going to change the value of the element in position 0, 0, which is the first element in the row with index 0 and column with index 0 to –1, we then run the following code:

```
two_dim[0,0] = -1
two_dim
```

This will change the value of the element from 3 to –1.

Slicing arrays

Slicing is getting or setting a smaller subarray within a larger array. You get slices from arrays in the same way you get slices from Python lists. If you're familiar with slicing Python lists, slicing arrays in NumPy is exactly the same. To understand better, let's reiterate on the array that we were using earlier by running the `one_dim` code. So, if you want all the elements from index 0 to index 5, not including index 5, you will have to run the cell with `print(one_dim[2,5])`. Likewise, if you want the first five elements, you can use `print(one_dim[:5])`, the element with index 5 will not be included. Again, just as with the Python lists, you can use negative indices. So, if you want to get the last five elements, you can use `print(one_dim[-5:])`:

```
In [33]:  two_dim = np.array([[3, 5, 2, 4], [7, 6, 5, 5], [1, 6, -1, -1]])
          two_dim

Out[33]:  array([[ 3,  5,  2,  4],
                 [ 7,  6,  5,  5],
                 [ 1,  6, -1, -1]])

In [34]:  two_dim[0,3]

Out[34]:  4

In [35]:  two_dim[0,0] = -1
          two_dim

Out[35]:  array([[-1,  5,  2,  4],
                 [ 7,  6,  5,  5],
                 [ 1,  6, -1, -1]])
```

On running the code, we get the preceding output. In the case of two-dimensional arrays, the rules are the same, but you need to specify one slice per dimension. Again, let's reiterate the two-dimensional arrays from our earlier example by using two_dim. In this case, if you want to get the initial four elements that are in the array, use two_dim[:2, :2]. Likewise, if you want to include the initial two elements from all the rows, use two_dim[:, 1:3] as shown in the following screenshot:

```
In [36]:  one_dim

Out[36]:  array([ 1. , -0.4, -0.3, -0.2, -0.1,  0. ,  0.1,  0.2,  0.3,  0.4,  0.5,
                  0.6])

In [37]:  print(one_dim[2:5])
          print(one_dim[:5])
          print(one_dim[-5:])

          [-0.3 -0.2 -0.1]
          [ 1.  -0.4 -0.3 -0.2 -0.1]
          [0.2 0.3 0.4 0.5 0.6]
```

In the preceding screenshot, we get the rows from the beginning up to the row with index 2, and the same thing for the columns. Similarly, we get the first two elements from all the rows of the array.

Reshaping arrays

Reshaping arrays changes an array from one dimension to another dimension. For example, one-dimension to two-dimension, one-dimension to three-dimension, and three-dimension to two-dimension. Let's now learn how to reshape arrays. Again, we will need to use the previous one-dimensional array. If you want to reshape or transform the array into a *4 x 3* two-dimensional array, do so using the `reshape()` method, `one_dim.reshape(4,3)` as shown in the following screenshot:

```
In [38]:  two_dim

Out[38]:  array([[-1,  5,  2,  4],
                 [ 7,  6,  5,  5],
                 [ 1,  6, -1, -1]])


In [39]:  two_dim[:2,:2]

Out[39]:  array([[-1,  5],
                 [ 7,  6]])


In [40]:  two_dim[:,1:3]

Out[40]:  array([[ 5,  2],
                 [ 6,  5],
                 [ 6, -1]])
```

On running the code, we get the preceding output; the array has been changed from a one-dimensional array with 12 elements to a two-dimensional array with four rows and three columns. You can also specify other dimensions if you want, such as 2 x 6, or you can even go to three dimensions, such as 2 x 2 x 3.

In this book, we will work only with one-dimensional and two-dimensional arrays, so don't worry about higher dimensions.

In the case of two-dimensional arrays, if you have a two-dimensional array and you want to convert it into a one-dimensional array, you can use the `flatten()` method as shown in the following screenshot:

```
In [61]:  one_dim

Out[61]:  array([-0.5, -0.4, -0.3, -0.2, -0.1,  0. ,  0.1,  0.2,  0.3,  0.4,  0.5,
                  0.6])

In [62]:  one_dim.reshape(4,3)

Out[62]:  array([[-0.5, -0.4, -0.3],
                 [-0.2, -0.1,  0. ],
                 [ 0.1,  0.2,  0.3],
                 [ 0.4,  0.5,  0.6]])
```

As shown in the preceding screenshot, on running the `flatten()` method, the two-dimensional array is transformed into a one-dimensional array.

Using NumPy for simulations

Now let's learn how to use NumPy in a real-world scenario. Here, we will cover two examples of simulations using NumPy, and in the process, we will also learn about other operations that we can do with arrays.

Coin flips

We will look into a coin flip, or coin toss, simulation using NumPy. For this purpose, we will use the `randint` function that comes in the random submodule from NumPy. This function takes the `low`, `high`, and `size` arguments, which will be the range of random integers that we want for the output. So, in this case, we want the output to be either 0 or 1, so the value for `low` will be 0 and `high` will be 2 but not including 2. Here, the `size` argument will define the number of random integers we want for the output, that is, the number of coins we will flip, in our case:

```
In [43]:  two_dim

Out[43]:  array([[-1,  5,  2,  4],
                 [ 7,  6,  5,  5],
                 [ 1,  6, -1, -1]])

In [44]:  two_dim.flatten()

Out[44]:  array([-1,  5,  2,  4,  7,  6,  5,  5,  1,  6, -1, -1])
```

So we will assign 0 as tails and 1 as heads and the `size` argument is assigned 1, since we will be flipping one coin. We will get a different result every time we run this simulation.

Let's take another simulation where you want to throw 10 coins at a time. Here, all you have to do is change the value of the last argument to `size=10`. And to get the total number of heads, you have to sum all the elements in the `experiment` output array:

```
In [45]:  np.random.randint(low=0, high=2, size=1)

Out[45]:  array([0])
```

Like the previous simulation, we will get random output every time we run the simulation. If you want to perform this experiment, many times, say 10,000 times, you can do it very easily using NumPy. Let's create a `coin_matrix` simulation to find out the distribution of the number of heads when throwing 10 coins at a time, we will use the same function, `randint`, with the same arguments, 0 and 2, but this time we want the size to be a two-dimensional array, so we will assign the `size=(10000,10)` argument. But since here we can't view the 10,000 rows-matrix on the screen, let's create a smaller matrix to display on the `coin_matrix[:5, :]` output:

```
In [45]:  np.random.randint(low=0, high=2, size=1)

Out[45]:  array([0])


In [46]:  experiment = np.random.randint(0,2, size=10)
          print(experiment)
          print(experiment.sum())

          [0 0 1 0 1 0 0 0 1 1]
          4
```

When we run the cell, we will get the first five rows of the matrix, and the result will be different every time we run the simulation.

 Here, the first five rows are the first five results of the 10 coins that we flip out of the 10,000 results that are in the actual matrix with 10,000 rows.

To calculate how many heads we got in every experiment, we can use the sum attribute, but in this case, we want to sum all the rows. To sum all the rows in NumPy, we use the additional arguments, axis and set axis=1; this will give you an array with a count of how many heads you get in every experiment:

```
In [48]:  counts = coin_matrix.sum(axis=1)
          print(counts[:25])
          print(counts.mean())
          print(np.median(counts))
          print(counts.min(), counts.max())
          print(counts.std())

          [5 4 7 8 6 2 7 5 6 8 6 4 6 6 2 5 6 5 6 3 3 4 6 7 6]
          4.9893
          5.0
          0 10
          1.5791724130062557
```

In the preceding screenshot, we called for the first 25 elements in the array, which contain the number of heads in every experiment. NumPy also provides arrays with some useful methods for performing statistics, such as mean, median, minimum, and maximum, and standard deviation. Using the mean() method, we will get the mean or the average of heads in all the experiments. The median() method will give us the median value for the total of heads from the experiments. You can use the min() and max() methods to get the minimum and maximum number of heads that we can get in our experiment. The std() method will calculate the standard deviation of the array counts.

 The output details for this section will be different every time we run the experiment. So do not be dismayed if your output doesn't match those mentioned earlier.

Now, if you want to know the distribution of the number of heads we get in the experiment, you can use the bincount function. If you run the cell, you will get an array of numbers that gives the number of heads for the experiments, starting from 0 to 10 as shown in the following screenshot:

```
In [48]: counts = coin_matrix.sum(axis=1)
         print(counts[:25])
         print(counts.mean())
         print(np.median(counts))
         print(counts.min(), counts.max())
         print(counts.std())

         [5 4 7 8 6 2 7 5 6 8 6 4 6 6 2 5 6 5 6 3 3 4 6 7 6]
         4.9893
         5.0
         0 10
         1.5791724130062557
```

The following code is just regular Python code that gives a detailed overview of the values for the distribution of the number of heads that you get in the experiment:

```
In [49]: np.bincount(counts)

Out[49]: array([  10,  109,  428, 1194, 2051, 2427, 2097, 1157,  431,   82,   14],
                dtype=int64)
```

The preceding screenshot depicts the details of the experiment that ran earlier; you can see that we got 0 heads 10 times, 1 heads 109 times, and so on, and also the percentages.

Simulating stock returns

Now let's look into another simulation example from the field of finance using the matplotlib NumPy library. Let's say we want to model the returns for a stock with the normal distribution. So, here we can use the normal function to produce random numbers that are normally distributed. In the normal function, we have the loc parameter, the scale parameter, also known as the standard deviation, and the parameter that holds the value of random numbers that we want. Here, the random parameter is the number of days in a trading year:

```
In [50]:   unique_numbers = np.arange(0,11)
           observed_times = np.bincount(counts)
           print("================\n")
           for n, count in zip(unique_numbers, observed_times):
               print("{} heads observed {} times ({:0.1f}%)".format(n, count, 100*count/2000

           ================

           0 heads observed 10 times (0.5%)
           1 heads observed 109 times (5.5%)
           2 heads observed 428 times (21.4%)
           3 heads observed 1194 times (59.7%)
           4 heads observed 2051 times (102.5%)
           5 heads observed 2427 times (121.3%)
           6 heads observed 2097 times (104.8%)
           7 heads observed 1157 times (57.9%)
           8 heads observed 431 times (21.6%)
           9 heads observed 82 times (4.1%)
           10 heads observed 14 times (0.7%)
```

When you run the cell or the simulation, you get an array of values that is the returns for the first 20 days. You will also get some negative and some positive returns, just like in normal stocks. Now let's say that you have initial_price as 100; and to calculate all the prices for all the following days you can apply initial_price times the exponential function of the cumulative sum of the returns.

Here, you must have a little background in finance to really understand this. The goal is not for you to understand how to perform simulations in finance, but to show you how easy it is to perform simulations using NumPy, and these are examples of how to use the simulations. Also, plots will be covered more elaborately in Chapter 4, *Visualization and Explanatory Data Analysis*, including how to perform them.

Now we will do some plots that will project the simulation of the stock using NumPy:

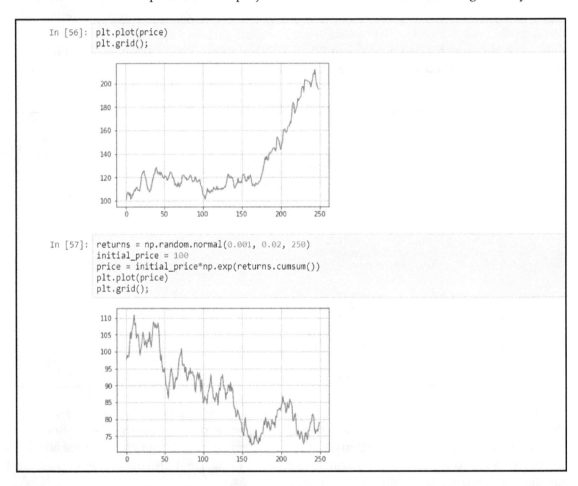

In the preceding screenshot, the stock is started at a price of **100**, and the evolution is plotted in the simulation. We have the same code, everything in one cell, and every time we run the cell, we will get a different simulation.

Summary

In this chapter, we learned about `numpy`, a library designed to do vectorized operations. We also learned about NumPy arrays, which are the main objects in NumPy. We learned how to create them, looked into their various attributes, explored which arrays are used in basic math, and did some manipulation with arrays. Then, we learned how to perform and run simple simulations using NumPy.

In the next chapter, we will look at `pandas`, the most popular library for doing data analysis in Python.

Pandas - Everyone's Favorite Data Analysis Library

3

In this chapter, we will introduce the pandas library. We will talk about Pandas, its capabilities, and its importance in the Python data science stack. We will also talk about series and DataFrames, the main objects in the pandas library. We will discuss their properties, and some of the operations and manipulations that we can do with these objects when doing data analysis. Further on, we will see some examples of how to use the objects in this library using a real-world dataset and answer some simple questions about datasets. The chapter will cover the following topics:

- The pandas library
- The operation and manipulation of pandas
- Answering simple questions about datasets with examples

Introduction to the pandas library

Pandas is a library for Python that provides fast, flexible, and expressive data structures designed to work with relational or tabular data, such as an SQL table, or an Excel spreadsheet. It is a fundamental, high-level building block for doing practical, real-world data analysis with Python. We use the following line of code to import the pandas library:

```
#The importing convention
import pandas as pd
```

Pandas is well-suited to the following cases:

- When you have tabular data with heterogeneously typed columns, such as the data that you can find in an SQL table or in an Excel spreadsheet
- When you have ordered or unordered time series data
- When you have data that is in rows and columns, similar to a matrix
- When you use, in your work, observational or other types of statistical datasets

There are two primary data structures in `pandas`:

- The series, which is a one-dimensional data structure
- The DataFrame, which is a two-dimensional data structure

They can handle the vast majority of cases for data analysis that you find in different fields, such as finance, statistics, social science, and many areas of engineering and business as well. Pandas is built on top of NumPy and it is intended to integrate well within a scientific computing environment with many other third-party libraries. You can also use `pandas` in combination with other libraries, and we will use it in this chapter with visualization libraries. Here are a few things that you can do with `pandas`:

- Easy handling of missing data in floating as well as non-floating point data
- Easy insertion and deletion of data in DataFrames and higher dimensional objects
- Automatic data alignment
- Group by functionality for powerful and flexible aggregation and the transformation of data
- Easy conversion of other differently indexed Python and NumPy data structures into DataFrame objects
- Intelligent label-based slicing, fancy indexing, and the subsetting of large datasets
- Intuitive merging and the joining of datasets
- Hierarchical labeling of axes
- Robust IO tools for loading data from flat files, Excel files, databases, and saving/loading data from the ultra-fast HDF5 format
- Time series specific functionality for date range generation and frequency conversion, moving window statistics, moving window linear regressions, date shifting, lagging, and so on

Important objects in pandas

The two most important objects in `pandas` are:

- Series
- DataFrames

To start using `pandas` in our work on data science with Python, let's first import the NumPy and math libraries with the following line of code:

```
import numpy as np
import matplotlib.pyplot as plt
%maplotlib inline
```

Then, to work with `pandas`, we import the `pandas` library using the standard convention: `import pandas as pd`.

Series

The series data structure in `pandas` is a one-dimensional labeled array. The characteristics of this data structure are:

- Data in a `pandas` series can be of any type, such as integers, strings, floating-point numbers, Python objects, and so on
- Data is homogeneous in nature or all the data must be of the same type

- Data always has an index that gives the following data structure dictionary and Python list or NumPy array type properties:

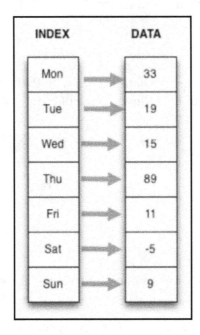

The preceding diagram is a visual example of a `pandas` series. You can see that every data point is associated with an index.

Creating a pandas series

There are many ways to create `pandas` series objects. The following are some of the most common ways:

- Creation from a list
- Creation from a dictionary
- Creation from a NumPy array
- Creation from an external data source, such as a file

Lets now learn how to create series from lists, dictionaries, and a NumPy array. First, let's define the data and index them as lists. Let's create a list of values and name it `temperature`, and another list of values named `days`. To create a series from the data, all you have to do is use the `pd.Series(temperature, index=days)` constructor:

```
In [3]:  # define the data and index as lists
         temperature = [33, 19, 15, 89, 11, -5, 9]
         days = ['Mon','Tue','Wed','Thu','Fri','Sat','Sun']

         # create series
         series_from_list = pd.Series(temperature, index=days)
         series_from_list
```

```
Out[3]:  Mon      33
         Tue      19
         Wed      15
         Thu      89
         Fri      11
         Sat      -5
         Sun       9
         dtype: int64
```

When we run the code, we see that each value is associated with the respective index. Next, we will create a series from a Python dictionary. In Python dictionaries, you always have keys associated with respective values. So, when you create a pandas series from a dictionary, all the keys will be used as the indices, and the respective values will be the values in the series associated with the index. Using the same data, from the preceding screenshot, we define our dictionary, my_dict, where the days are associated with each temperature. Then, we pass the dictionary to thepd.Series(my_dict) constructor, which will give us the following result:

```
In [4]:  # from a dictionary
         my_dict = {'Mon': 33, 'Tue': 19, 'Wed': 15, 'Thu': 89, 'Fri': 11, 'Sat': -5, 'Sun': 9}
         series_from_dict = pd.Series(my_dict)
         series_from_dict
```

```
Out[4]:  Mon      33
         Tue      19
         Wed      15
         Thu      89
         Fri      11
         Sat      -5
         Sun       9
         dtype: int64
```

Although the days of the week are not ordered, because we don't have an implicit order in Python dictionaries, each day is associated with its own temperature. For example `Friday` is associated with `11`; `Monday` is associated with `33`, and so on.

Next, we will look at how to create a `pandas` series from a NumPy array. First, let's define an object, `my_array`, using the `np.linspace` function. Next, we pass the object to the `pd.Series` constructor, which will create a series from the defined NumPy array, as seen in the following screenshot:

```
In [5]:  # From a numpy array
         my_array = np.linspace(0,10,15)
         series_from_ndarray = pd.Series(my_array)
         series_from_ndarray

Out[5]:  0      0.000000
         1      0.714286
         2      1.428571
         3      2.142857
         4      2.857143
         5      3.571429
         6      4.285714
         7      5.000000
         8      5.714286
         9      6.428571
         10     7.142857
         11     7.857143
         12     8.571429
         13     9.285714
         14    10.000000
         dtype: float64
```

Since we didn't specify any index, `pandas` created an automatic integer index that starts with `0` and goes up to the number of elements that we have as -1. So, in this case, we have `15` elements with an index that goes to `14`.

You can also do vectorized operations with `pandas` series, similar to how you can with NumPy arrays. If you do one operation on a series, the same operation will be applied to every element in the series. The following screenshot shows a few examples of this:

```
In [59]: #Vectorized operations also work in pandas Series
         2*series_from_list

Out[59]: Mon      66
         Tue      38
         Wed      30
         Thu     178
         Fri      22
         Sat     -10
         Sun      18
         dtype: int64


In [60]: #Vectorized operations also work in pandas Series
         series_from_list + 2

Out[60]: Mon      35
         Tue      21
         Wed      17
         Thu      91
         Fri      13
         Sat      -3
         Sun      11
         dtype: int64


In [61]: #Vectorized operations also work in pandas Series
         np.exp(series_from_list)

Out[61]: Mon     2.146436e+14
         Tue     1.784823e+08
         Wed     3.269017e+06
         Thu     4.489613e+38
         Fri     5.987414e+04
         Sat     6.737947e-03
         Sun     8.103084e+03
         dtype: float64
```

In the preceding screenshot, we have multiplied the series by 2, where every element gets multiplied by 2. Again, we added 2 to the series and we see that every element gets added by 2. Similarly, you can do other types of arithmetic operations and you can even apply universal functions from NumPy. For this, we have calculated the exponential of the series using the NumPy, np.exp mathematical function. On running this operation, we see another pandas series with the same index and with the new exponential values of the original series.

DataFrames

A DataFrame is a two-dimensional labeled data structure with columns of potentially different types. A `pandas` DataFrame is similar to an Microsoft Excel spreadsheet, or an SQL table. You have two indices, the index for the rows and the index for the columns:

INDEX	Dates	Tokyo	Paris	Mumbai
0	12-1	15	-2	20
1	12-2	19	0	18
2	12-3	15	2	23
3	12-4	11	5	19
4	12-5	9	7	25
5	12-6	8	-5	27
6	12-7	13	-3	23

In the preceding diagram, we have two indices, columns associated with values of **Dates**, **Tokyo**, and so on, and a row associated with **INDEX**.

Creating a pandas DataFrame

There are many ways to create `pandas` DataFrames, but the most important and commonly used method is when you create `pandas` DataFrames from a file. You can create DataFrames from the following:

- Dict of 1-D `ndarrays`, list, dicts, or series
- 2-D `numpy.ndarray`
- The TEXT, CSV, Excel files, or databases

Let's create a DataFrame from a real-world dataset. The dataset that we will use is one that contains data about human resources, employee attrition, performance, and so on. You can get it from the following link: `https://www.ibm.com/communities/analytics/watson-analytics-blog/hr-employee-attrition/`.

> One very interesting thing about pandas is that you can download data directly from the internet. If the file is in some URL, you can save the URL in a Python string and create a DataFrame directly using the `read_excel` function from `pandas`.

Since we will be creating the DataFrame from an Excel file, we will use just a few arguments. The first argument will be the `io` argument, which specifies the location of the file. Then we have the `sheetname`, which defines the sheet you want to read from the Excel file. The third and final argument will be the `index_col` that you want to use as an index:

```
In [7]:  file_url = "https://community.watsonanalytics.com/wp-content/uploads/2015/03/WA_F

In [8]:  data = pd.read_excel(io=file_url, sheetname=0, index_col='EmployeeNumber')
```

So, in the preceding screenshot, we had our `file_url` variable defined, which is the file location for the following cell. We then created the DataFrame and named it `data`.

Anatomy of a DataFrame

A DataFrame consists of three parts:

- An index
- The column names
- The data

The row and column labels can be accessed respectively by accessing the `index` and `columns` attributes. This is showcased in the following screenshot:

```
In [9]:   data.columns

Out[9]:   Index(['Age', 'Attrition', 'BusinessTravel', 'DailyRate', 'Department',
                 'DistanceFromHome', 'Education', 'EducationField', 'EmployeeCount',
                 'EnvironmentSatisfaction', 'Gender', 'HourlyRate', 'JobInvolvement',
                 'JobLevel', 'JobRole', 'JobSatisfaction', 'MaritalStatus',
                 'MonthlyIncome', 'MonthlyRate', 'NumCompaniesWorked', 'Over18',
                 'OverTime', 'PercentSalaryHike', 'PerformanceRating',
                 'RelationshipSatisfaction', 'StandardHours', 'StockOptionLevel',
                 'TotalWorkingYears', 'TrainingTimesLastYear', 'WorkLifeBalance',
                 'YearsAtCompany', 'YearsInCurrentRole', 'YearsSinceLastPromotion',
                 'YearsWithCurrManager'],
                dtype='object')


In [10]:  data.index

Out[10]:  Int64Index([   1,    2,    4,    5,    7,    8,   10,   11,   12,   13,
                       ...
                      2054, 2055, 2056, 2057, 2060, 2061, 2062, 2064, 2065, 2068],
                     dtype='int64', name='EmployeeNumber', length=1470)


In [11]:  data.values

Out[11]:  array([[41, 'Yes', 'Travel_Rarely', ..., 4, 0, 5],
                 [49, 'No', 'Travel_Frequently', ..., 7, 1, 7],
                 [37, 'Yes', 'Travel_Rarely', ..., 0, 0, 0],
                 ...,
                 [27, 'No', 'Travel_Rarely', ..., 2, 0, 3],
                 [49, 'No', 'Travel_Frequently', ..., 6, 0, 8],
                 [34, 'No', 'Travel_Rarely', ..., 3, 1, 2]], dtype=object)
```

Using the `columns` attribute will give all the column names and, similarly, using the `index` attribute will give the index. Finally, using the `value` attribute will give the values of the DataFrame, which will always be in the form of a NumPy array.

Operations and manipulations of pandas

There are a number of operation methods used to work in `pandas`. In this section of the chapter, we will look into some of the common operations that we will be doing in this book.

Inspection of data

The first thing that you will want to do when loading a DataFrame or creating a DataFrame from a file is to inspect the data that you just loaded. We have two methods for inspecting the data:

- Head
- Tail

The `head` method will show us the data of the first five rows:

```
In [12]: data.head()
Out[12]:
```

EmployeeNumber	Age	Attrition	BusinessTravel	DailyRate	Department	DistanceFromHome	Ed(
1	41	Yes	Travel_Rarely	1102	Sales	1	
2	49	No	Travel_Frequently	279	Research & Development	8	
4	37	Yes	Travel_Rarely	1373	Research & Development	2	
5	33	No	Travel_Frequently	1392	Research & Development	3	
7	27	No	Travel_Rarely	591	Research & Development	2	

5 rows × 34 columns

As you can see in the preceding screenshot, we have the first five rows and 34 columns displayed on running the `data.head()` method:

```
In [13]:  data.tail()
Out[13]:
```

	Age	Attrition	BusinessTravel	DailyRate	Department	DistanceFromHome	Edu
EmployeeNumber							
2061	36	No	Travel_Frequently	884	Research & Development	23	
2062	39	No	Travel_Rarely	613	Research & Development	6	
2064	27	No	Travel_Rarely	155	Research & Development	4	
2065	49	No	Travel_Frequently	1023	Sales	2	
2068	34	No	Travel_Rarely	628	Research & Development	8	

5 rows × 34 columns

To take a look at the last five rows of data, you use the `tail` method. The preceding screenshot shows us the last five rows of data from our DataFrame on running the `data.tail()` method. It is always a good practice to use these two methods to make sure that the data is correctly loaded.

Selection, addition, and deletion of data

You can view DataFrames as a dictionary of a series, where every column will be similar to a `pandas` series and you can access this series just as the same as how you can access objects in a dictionary. You can treat a DataFrame as a dictionary of indexed series objects. Getting, setting, and deleting columns works with the same syntax as you work with dictionaries. Let's look at a few examples. Suppose you want to access the Age column in your DataFrame. All you have to do is to write the method and indicate the name of the column you want to access:

```
In [14]:   # Getting one column: .head() is just to print the first 5 values
           data['Age'].head()

Out[14]:   EmployeeNumber
           1     41
           2     49
           4     37
           5     33
           7     27
           Name: Age, dtype: int64

In [15]:   # Getting more than one column
           data[['Age', 'Gender','YearsAtCompany']].head()

Out[15]:
```

	Age	Gender	YearsAtCompany
EmployeeNumber			
1	41	Female	6
2	49	Male	10
4	37	Male	0
5	33	Female	8
7	27	Male	2

In the preceding screenshot, when we executed the `data['Age'].head()` and
`data[['Age', 'Gender', 'YearsAtCompany']].head()` methods, we got the first five
rows from each of the mentioned columns from the DataFrame.

> Note that `.head()` is used to avoid displaying all the rows in the
> DataFrame, which would be too long a list.

To add a column to a DataFrame, you create the column using the `data['AgeInMonths']` `= 12*data['Age']` method:

```
In [16]:  # Adding a column
          data['AgeInMonths'] = 12*data['Age']
          data['AgeInMonths'].head()

Out[16]:  EmployeeNumber
          1      492
          2      588
          4      444
          5      396
          7      324
          Name: AgeInMonths, dtype: int64
```

This column doesn't currently exist in the DataFrame, but by executing the method mentioned just now, a new column, AgeInMonths, will be added to the DataFrame.

You can use the `del` statement to delete the column that we just created in the DataFrame. Another way to delete a column from a DataFrame is to use the `drop` method. To use the `drop` method, you pass the column that you want to delete, and if you are deleting columns, you should specify that this will be from `axis=1`, and the other argument `inplace` means that you want to modify the data object in place, so you want this change to be permanent in the data structure. These operations are shown in the following screenshot:

```
In [17]:  # Deleting a column
          del data['AgeInMonths']
          # the drop method can also be used
          data.drop('EmployeeCount', axis=1, inplace=True)

In [18]:  data.columns

Out[18]:  Index(['Age', 'Attrition', 'BusinessTravel', 'DailyRate', 'Department',
                 'DistanceFromHome', 'Education', 'EducationField',
                 'EnvironmentSatisfaction', 'Gender', 'HourlyRate', 'JobInvolvement',
                 'JobLevel', 'JobRole', 'JobSatisfaction', 'MaritalStatus',
                 'MonthlyIncome', 'MonthlyRate', 'NumCompaniesWorked', 'Over18',
                 'OverTime', 'PercentSalaryHike', 'PerformanceRating',
                 'RelationshipSatisfaction', 'StandardHours', 'StockOptionLevel',
                 'TotalWorkingYears', 'TrainingTimesLastYear', 'WorkLifeBalance',
                 'YearsAtCompany', 'YearsInCurrentRole', 'YearsSinceLastPromotion',
                 'YearsWithCurrManager'],
                dtype='object')
```

Now, if we take a look at the columns of data, which is our DataFrame, we see that
`EmployeeCount` has been deleted.

Slicing DataFrames

Just as you can do with NumPy series, you can take slices from both `pandas` series and
DataFrames. We can use the same notation in series and DataFrames for slicing:

```
In [19]: data['BusinessTravel'][10:15]

Out[19]: EmployeeNumber
         14     Travel_Rarely
         15     Travel_Rarely
         16     Travel_Rarely
         18     Travel_Rarely
         19     Travel_Rarely
         Name: BusinessTravel, dtype: object

In [20]: data[10:15]

Out[20]:
```

EmployeeNumber	Age	Attrition	Business Travel	DailyRate	Department	DistanceFromHome	Edu
14	35	No	Travel_Rarely	809	Research & Development	16	
15	29	No	Travel_Rarely	153	Research & Development	15	
16	31	No	Travel_Rarely	670	Research & Development	26	
18	34	No	Travel_Rarely	1346	Research & Development	19	
19	28	Yes	Travel_Rarely	103	Research & Development	24	

5 rows × 33 columns

In the preceding screenshot, we have an example of slicing from a and a DataFrame. The
data from position 10 to 15 (excluding 15) is sliced and displayed in the output.

Selection by labels

You can also perform selection by labels, and this is why the index is very important in the data structures. If you want the data from some specific employees, you can get them by using the `loc` method. Let's first specify the employees for whom we want the data by defining `selected_EmployeeNumbers = [15, 94, 337, 1120]`. Since in both of the data structures every value is associated with the `EmployeeNumber` index, we can use this index to access the data that we want specifically:

```
In [21]: selected_EmployeeNumbers = [15, 94, 337, 1120]

In [22]: data['YearsAtCompany'].loc[selected_EmployeeNumbers]
Out[22]: EmployeeNumber
         15       9
         94       5
         337      2
         1120     7
         Name: YearsAtCompany, dtype: int64
```

So, in the preceding screenshot, we have the data from a `pandas` series that contains the number of years at the company.

The `loc` method is also available in the DataFrame. If you pass the same list to the `loc` method in the DataFrame, you will get all the data associated with the labels. This is shown in the following screenshot:

```
In [23]:  data.loc[selected_EmployeeNumbers]
Out[23]:
```

	Age	Attrition	BusinessTravel	DailyRate	Department	DistanceFromHome	Edi
EmployeeNumber							
15	29	No	Travel_Rarely	153	Research & Development	15	
94	29	No	Travel_Rarely	1328	Research & Development	2	
337	31	No	Travel_Frequently	1327	Research & Development	3	
1120	29	No	Travel_Rarely	1107	Research & Development	28	

4 rows × 33 columns

```
In [24]:  # Getting a single value
          data.loc[94,'YearsAtCompany']
Out[24]:  5
```

If you want to get a specific value or a specific cell from your DataFrame, you must pass two indices, the index for the row and the index for the column, to the `loc` method. You can also access data by position using the `iloc` method.

Answering simple questions about a dataset

Let's take an example to look at a few questions and answer them. Say the HR director asks you to answer a few descriptive questions about the employees of the company. The following are a few such questions:

- How many employees are there by department in the dataset?
- What is the overall attrition rate?
- What is the average hourly rate?
- What is the average number of years at the company?
- Who are the five employees with the most number of years at the company?
- How satisfied are employees overall?

Total employees by department in the dataset

To view the departments in the dataset, you use the `data['Department']` statement. We get the column called `Department`, which is a `pandas` series, and for every employee, you have the department to which the employee belongs. So, to compute how many times you see each department in this `pandas` series, you use the `value_counts()` method:

```
In [25]:  data['Department'].value_counts()

Out[25]:  Research & Development    961
          Sales                     446
          Human Resources            63
          Name: Department, dtype: int64
```

In the preceding screenshot, you can see that the count for every unique value from `Department` that you have in your `pandas` series has been tabulated and displayed.

Overall attrition rate

First, let's use the `data['Attrition']` statement to view the `Attrition` column in the dataset. In the column, we see the data that shows whether the employee is still at the company. For the employees that are still at the company, we have `Attrition` equal to `No`, and for employees that have left the company, we have `Attrition` equal to `Yes`. We then pass the `value_counts` method to compute the occurrence of each conclusion. To get the attrition rate, which is the proportion of employees that have left the company, we can use the additional argument, `normalize=True`. The operations mentioned here are shown in the following screenshot:

```
In [26]:  data['Attrition'].value_counts()

Out[26]:  No     1233
          Yes     237
          Name: Attrition, dtype: int64

In [27]:  data['Attrition'].value_counts(normalize=True)

Out[27]:  No     0.838776
          Yes    0.161224
          Name: Attrition, dtype: float64

In [28]:  attrition_rate = data['Attrition'].value_counts(normalize=True)['Yes']
          attrition_rate

Out[28]:  0.16122448979591836
```

Now, to get the overall attrition rate, we use the associated label in the `Yes` index. So, in the preceding screenshot, we get the value of the `Yes` index, which is an attrition rate of 16.12%.

Average hourly rate

There are a lot of statistical methods that you can use in `pandas` series. One of the statistical methods that you will use all the time is the `mean` method, which computes the mean of your `pandas` series:

```
In [29]:  data['HourlyRate'].mean()

Out[29]:  65.89115646258503
```

So, by using this method, we find that the mean for the `HourlyRate` variable is `65.89`.

Average number of years

To get the average number of years at the company, you can use the `mean` method.
However, there is another very useful method by which you get not only the mean, but also
many other descriptive statistics of the series. This method is known as the `describe()`
method, which gives the mean, the standard deviation, the minimum, the maximum, and
the percentiles:

```
In [30]:   data['YearsAtCompany'].describe()

Out[30]:   count    1470.000000
           mean        7.008163
           std         6.126525
           min         0.000000
           25%         3.000000
           50%         5.000000
           75%         9.000000
           max        40.000000
           Name: YearsAtCompany, dtype: float64
```

In the preceding screenshot, we see the average number of years at the company is 7. So,
this is another very handy method that you can always use.

Employees with the most number of years

There is another method that you can use for sorting the values in a pandas Series. The
`sort_values` method will sort the values, and by default, it will sort them in ascending
order. If you don't want them in ascending order, you can specify this argument to be
`False`. As always, it will display the whole list of data, so we have used the `slice` notation
to get the first five elements in the Series:

```
In [31]:   data['YearsAtCompany'].sort_values(ascending=False)[:5]

Out[31]:   EmployeeNumber
           165      40
           131      37
           374      36
           1578     36
           776      34
           Name: YearsAtCompany, dtype: int64


In [32]:   data['YearsAtCompany'].sort_values(ascending=False)[:5].index

Out[32]:   Int64Index([165, 131, 374, 1578, 776], dtype='int64', name='EmployeeNumber')
```

In the preceding screenshot, we use the `index` method to get the employee numbers that identify the employees with the most number of years at the company.

Overall employee satisfaction

In the dataset, we have a column called `JobSatisfaction`, which holds employee satisfaction ratings ranging from 1 to 4. Let's use the `head` method to see the first five observations and create a dictionary where the number 1 corresponds to low job satisfaction, the value 2 corresponds to medium job satisfaction, and so on. Now we will transform the code into the respective categories using the `map` method for the pandas Series. This will associate each key with the respective value, that is, 1 will be transformed into `Low`, 2 to `Medium`, and so on. Then, we will reassign the Series to the Series operated with the `map` method:

```
In [33]:   data['JobSatisfaction'].head()

Out[33]:   EmployeeNumber
           1    4
           2    2
           4    3
           5    3
           7    2
           Name: JobSatisfaction, dtype: int64


In [34]:   JobSatisfaction_cat = {
               1: 'Low',
               2: 'Medium',
               3: 'High',
               4: 'Very High'
           }

           Transform this encodings to meaninful labels

In [35]:   data['JobSatisfaction'] = data['JobSatisfaction'].map(JobSatisfaction_cat)
           data['JobSatisfaction'].head()

Out[35]:   EmployeeNumber
           1    Very High
           2       Medium
           4         High
           5         High
           7       Medium
           Name: JobSatisfaction, dtype: object
```

As you can see in the following screenshot, we have the categories mapped just as we wanted. So the `map` method is very useful whenever you want to perform these kinds of transformations in your pandas Series. Next, we use the `value_counts()` method to get the occurrence of each category. But because of the question that was asked, it might be more useful to get the normalized count. So, if you multiply the Series by `100`, you will get the percentages:

```
In [36]:   data['JobSatisfaction'].value_counts()

Out[36]:   Very High     459
           High          442
           Low           289
           Medium        280
           Name: JobSatisfaction, dtype: int64


In [37]:   100*data['JobSatisfaction'].value_counts(normalize=True)

Out[37]:   Very High     31.224490
           High          30.068027
           Low           19.659864
           Medium        19.047619
           Name: JobSatisfaction, dtype: float64
```

So, we can see in the screenshot that `31%` of the employees in this dataset have job satisfaction that is considered `Very High`, and you can also see the other corresponding percentages for the other categories.

Answering further questions

Let's say that you are done with the first round of questions that the HR director asked you, but now he wants to know a little more about the employees. The following are the new tasks assigned to you:

- Give me a list of the employees with a Low level of `JobSatisfaction`
- Give me a list of the employees with a Low level of both `JobSatisfaction` and `JobInvolvement`
- Compare the employees with Low and Very High `JobSatisfaction` across the following variables: `Age`, `Department`, and `DistanceFromHome`

Employees with Low JobSatisfaction

To answer this question, we use a Boolean Series to index a series or a DataFrame. This is called **masking**, or **Boolean indexing**. First, we use the comparison operator to compare the pandas Series with a `data['JobSatisfaction'] == 'Low'` value, as follows:

```
In [42]: data['JobSatisfaction'] == 'Low'

Out[42]: EmployeeNumber
         1          False
         2          False
         4          False
         5          False
         7          False
         8          False
         10          True
         11         False
         12         False
         13         False
         14         False
         15         False
         16         False
         18         False
         19         False
         20          True
         21         False
         22         False
         23         False
         24         False
         26         False
         27          True
         28         False
         30         False
         31          True
```

If you run the operation, you will get a Boolean Series with the values `True` or `False` for each employee. `True` is where the value of `JobSatisfaction` for an employee is equal to Low and `False` is when the values are not equal to Low.

You can use this pandas Boolean Series to index another object, such as a Series or a DataFrame. When you use a Boolean Series to index another object, you will get back a new Series or DataFrame with the observations where you have `True` values. Let's use the following feature to answer our question:

```
In [43]:  data.loc[data['JobSatisfaction'] == 'Low'].index

Out[43]:  Int64Index([  10,   20,   27,   31,   33,   38,   51,   52,   54,   68,
                        ...
                      1975, 1980, 1998, 2021, 2023, 2038, 2054, 2055, 2057, 2062],
                     dtype='int64', name='EmployeeNumber', length=289)
```

So, if we use this Boolean Series to index our DataFrame using the `data.loc[data['JobSatisfaction'] == 'Low'].index` index attribute, you get the list of employees with a Low level of `JobSatisfaction`.

Employees with both Low JobSatisfaction and JobInvolvement

The `JobInvolvement` column has the same properties as the `JobSatisfaction` column, so instead of the categories, we have the corresponding numbers. We will first apply the `map` transformations that we applied previously to the `JobSatisfaction` column:

```
In [44]:  JobInvolment_cat = {
                1: 'Low',
                2: 'Medium',
                3: 'High',
                4: 'Very High'
          }
          data['JobInvolvement'] = data['JobInvolvement'].map(JobInvolment_cat)
```

Since we want only the employees with a Low level in both of the columns, we can use an &
logical operator to run the operation on the two Boolean Series to get the list of employees
with a low level of both `JobSatisfaction` and `JobInvolvement`. Again, we will use the
`loc` selection method and the `index` attribute to get the specific requirements:

```
In [62]: loc[(data['JobSatisfaction'] == 'Low') & (data['JobInvolvement'] == 'Low')].index

Out[62]: Int64Index([33, 235, 454, 615, 1019, 1037, 1237, 1460, 1478, 1544, 1611, 1622,
                     1905, 1956],
                    dtype='int64', name='EmployeeNumber')
```

In the preceding screenshot, we have a list of the employee numbers with a Low level of
both `JobSatisfaction` and `JobInvolvement`.

Employee comparison

Now we will compare the employees with Low and Very High `JobSatisfaction` levels.
We will compare these two groups across the following variables: Age, Department, and
`DistanceFromHome`. To make this comparison, we will use a grouping operation.
Grouping operations are those operations where we apply a sequence of actions. These
actions are called split, apply, and combine. We will see how this works in the following
steps:

1. The first step in this operation is the **Split** step, where we split our DataFrame
   into a group of DataFrames based on some criteria or some categorical variable.
   This will produce a grouped object that has a structure similar to a dictionary,
   and every group is associated with a different key.
2. The second step is to apply a function to every part of this group-by/grouped
   object, so we can apply a function and get a result.
3. The third step in this operation is to group or combine the results of the previous
   operation into a new data structure. This new data structure can be a `new_df` or
   it can be a `new_series`.

Now, to make the comparison, let's create a new DataFrame that contains only those observations that are given as criteria for comparison and call the DataFrame `subset_of_interest`. We will use the `bitwise or |` operator to get the Series of employees with Low and High `JobSatisfaction` and combine them and use the `shape` attribute to see the shape of our newly created DataFrame. We then use the `value_count` operation to see the exact counts of each observation:

```
In [46]:  subset_of_interest = data.loc[(data['JobSatisfaction'] == "Low") | (data['JobSati
          subset_of_interest.shape

Out[46]:  (748, 33)

In [47]:  subset_of_interest['JobSatisfaction'].value_counts()

Out[47]:  Very High    459
          Low          289


          Name: JobSatisfaction, dtype: int64
```

In the preceding screenshot, we see that this DataFrame has `748` rows and `33` columns; also, we have `459` observations with Very High `JobSatisfaction`, and `289` with Low level of `JobSatisfaction`.

Now that we have only the observations that we are interested in, we can apply the `groupby` operations to our newly created DataFrame to compare across the variables we wanted. We will name the object that we will get back from this operation `grouped`:

```
In [50]:  grouped = subset_of_interest.groupby('JobSatisfaction')

In [51]:  grouped.groups

Out[51]:  {'Low': Int64Index([  10,    20,    27,    31,    33,    38,    51,    52,    54,    68,
                        ...
                    1975, 1980, 1998, 2021, 2023, 2038, 2054, 2055, 2057, 2062],
                   dtype='int64', name='EmployeeNumber', length=289),
           'Medium': Int64Index([], dtype='int64', name='EmployeeNumber'),
           'High': Int64Index([], dtype='int64', name='EmployeeNumber'),
           'Very High': Int64Index([   1,     8,    18,    22,    23,    24,    30,    36,    39,
              40,
                        ...
                    2022, 2024, 2027, 2036, 2040, 2041, 2045, 2052, 2056, 2061],
                   dtype='int64', name='EmployeeNumber', length=459)}
```

As you can see, applying the `groups` method to this newly created object, we have two groups, the Low group and the Very High group, and you get the corresponding labels, or the indices, for every associated group. If you want to actually get the data associated with every group, you can use the `get_group` method. This operation is shown in the following screenshot:

```
In [52]: grouped.get_group('Low').head()
Out[52]:
```

EmployeeNumber	Age	Attrition	BusinessTravel	DailyRate	Department	DistanceFromHome	Edu
10	59	No	Travel_Rarely	1324	Research & Development	3	
20	29	No	Travel_Rarely	1389	Research & Development	21	
27	36	Yes	Travel_Rarely	1218	Sales	9	
31	34	Yes	Travel_Rarely	699	Research & Development	6	
33	32	Yes	Travel_Frequently	1125	Research & Development	16	

5 rows × 33 columns

If you use the `get_group` method with the `Low` key, you will get the DataFrame associated with the Low group.

> An interesting thing about these grouped objects is that you can actually get the Series from the original DataFrame. But if you try to get the Series by name, you will not get an ordinary Series, but you will get what is called a `groupby` Series, and when you apply a method, for instance the `mean` method, you will see that this method will be applied across every group.

So, if you call the `Age` Series of the grouped object and then ask for the mean, you will get the mean for every group. You can also use every method from an ordinary Series in a `groupby` Series. If you use the `describe` method, you will see that the `describe` method will be applied to both the Low group and the Very High group:

```
In [51]: grouped['Age']

Out[51]: <pandas.core.groupby.SeriesGroupBy object at 0x000001FBEABD23C8>
```

```
In [52]: grouped['Age'].mean()

Out[52]: JobSatisfaction
         Low          36.916955
         Very High    36.795207
         Name: Age, dtype: float64
```

```
In [53]: grouped['Age'].describe()

Out[53]: JobSatisfaction
         Low              count    289.000000
                          mean      36.916955
                          std        9.245496
                          min       19.000000
                          25%       30.000000
                          50%       36.000000
                          75%       42.000000
                          max       60.000000
         Very High        count    459.000000
                          mean      36.795207
                          std        9.125609
                          min       18.000000
                          25%       30.000000
                          50%       35.000000
                          75%       43.000000
                          max       60.000000
         Name: Age, dtype: float64
```

Once we apply the methods to the new Series, we see that the mean age for the Low group is `36.9`, and for the Very High group it's `36.7`. We also have the output for the `describe` method that we applied to both the Low group and the Very High group. The Series that we get from the `describe` method is a pandas Series, but this Series has a multi-level index. The first level of the index is the Low and Very High groups, and the second level of the index is the name of the statistical value, such as `mean`, `std`, `min`, and so on:

```
In [54]: grouped['Age'].describe().unstack()
```

Out[54]:

	count	mean	std	min	25%	50%	75%	max
JobSatisfaction								
Low	289.0	36.916955	9.245496	19.0	30.0	36.0	42.0	60.0
Very High	459.0	36.795207	9.125609	18.0	30.0	35.0	43.0	60.0

You can use the unstack method to transform the result into a DataFrame, as seen in the preceding screenshot.

Next, to compare across departments, we can use the value_counts and unstack methods to see the number of people that are in every department for each of the groups that we are interested in, as follows:

```
In [55]: grouped['Department'].value_counts().unstack()
```

Out[55]:

Department	Human Resources	Research & Development	Sales
JobSatisfaction			
Low	11	192	86
Very High	17	295	147

```
In [56]: 100*grouped['Department'].value_counts(normalize=True).unstack()
```

Out[56]:

Department	Human Resources	Research & Development	Sales
JobSatisfaction			
Low	3.806228	66.435986	29.757785
Very High	3.703704	64.270153	32.026144

To make a comparison, we use the normalize operation, where we see that 3% of the people with a Low level of JobSatisfaction belong to the Human Resources department, 66% belong to the Research & Development department, and 29% to the Sales department, and, similarly, we have the details for the Very High group too.

Finally, for the `DistanceFromHome` comparison, we can use the `describe` method and the `unstack` method to transform it into a DataFrame and we can compare the two groups across the `DistanceFromHome` variable:

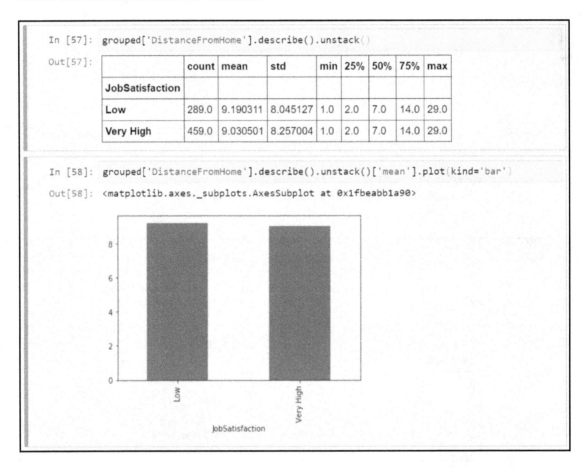

In the preceding screenshot, we have a DataFrame from the result of the `DistanceFromHome` comparison and also a bar graph plot from the mean of the resultant comparison.

Summary

In this chapter, we talked about pandas and its main objects, Series and DataFrames, and the main operations that we can perform with them. We also used pandas to answer questions similar to ones that every data analyst will come across a daily basis by doing some real-world data analysis and answering more complex questions.

In the next chapter, we will learn about visualization and exploratory data analysis with very powerful tools, such as matplotlib and the seaborn library.

Visualization and Exploratory Data Analysis

4

Visualization is a key topic for data science and data analysis, and Python provides a lot of options in terms of executing visualizations for different purposes. In this chapter, we will talk about the two most popular libraries for doing visualization in Python, namely, matplotlib and seaborn. We will also talk about the pandas capabilities for doing visualizations.

Let's look into the following various topics that we will discuss in this chapter:

- Introducing matplotlib
- Introducing pyplot
- Object-oriented interfaces
- Common customizations
- Exploratory data analysis with seaborn and pandas
- Analyzing the variables individually
- The relationship between variables

Introducing Matplotlib

Matplotlib tries to make easy things easier, and hard things possible. Basically, `matplotlib` is a plotting library that produces publication quality figures in a variety of formats and interactive environments. Let's now discuss what `matplotlib` is, its capabilities, and also its basic concepts, figures, subplots (axes), and axes. It can be used everywhere and for a variety of purposes. It can also be used in Python scripts, Python interpreter, the Python shell, the Jupyter Notebook, web application servers, and every graphical user interface that we can produce with Python.

Now, let's take a look at our Jupyter Notebook, wherein lies more information about `matplotlib`. But before doing that, let's first visit the website `matplotlib.org`. This is the project website and the primary online resource for this library's documentation. We can find examples, frequently asked questions, and the gallery, which is something we need to look at.

What people usually do when they work with `matplotlib` is they go to the gallery. We can see a representation of it a visualization that approximates to what they are trying to do in the following screenshot:

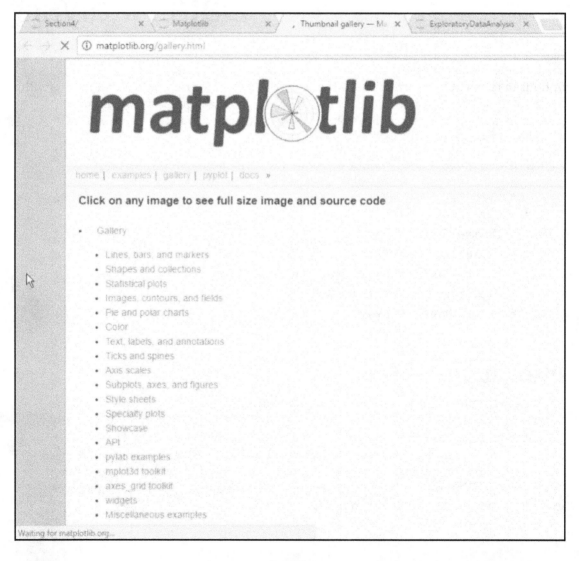

Let's say that we want to do a box plot. Let's see an example in the following screenshot. We see a visualization that compares a **violin plot** versus a **box plot** and when we have something similar to this in mind, we can look for the code, tweak it, and start using it for our own visualization:

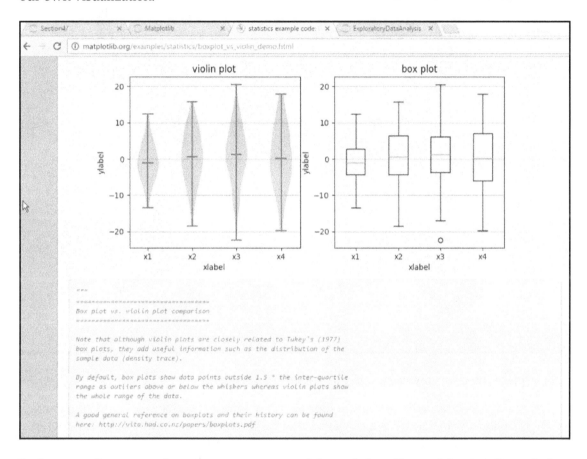

In the preceding screenshot, we can see a part of the code but, if we wish to go through the entire code, we can refer this code to the official `matplotlib` site mentioned earlier.

Terminologies in Matplotlib

Before talking about the main concepts of this library, we will discuss some basic terminologies that we have in `matplotlib`, such as figures, subplots/axes, and axis. The anatomy of a `matplotlib` plot starts with the figure that we can see in the following screenshot:

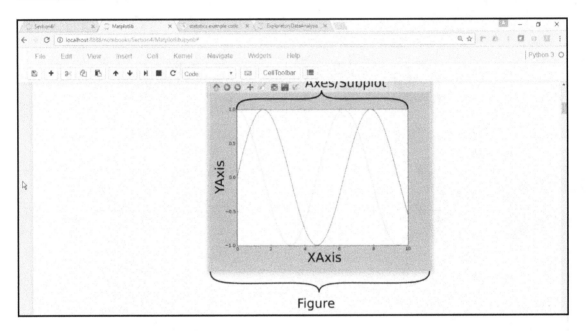

Let's explore the terms mentioned in the preceding screenshot:

- **Figure**: The figure is the first top-level container in this hierarchy. It is the overall window that contains everything that is drawn. We can have multiple independent figures and multiple axes in the figure.
- **Axes/Subplot**: Now, most of the plotting is done with respect to one axis or subplot. This plot has a lot of components to it, such as the *x* axis and the *y* axis; we have a plotting area, we have tick marks, and so on. As part of the subplot, we have other objects such as the *x* axis and, within the *x* axis, we have things such as the x label, the x tick marks, and the labels for the tick marks. This is basically the hierarchy that we have in `matplotlib`.

- **Axis:** We can see that the top of the hierarchy has the figure and, inside the figure, we have subplots. But the preceding image has only one subplot, but otherwise, we can have many subplots inside a figure. Every subplot has other elements; most commonly, we will have an *x* axis, a *y* axis, and many other elements.

Introduction to pyplot

Now, we will start using `matplotlib` with the `pyplot` interface. The topic that we will cover is the `pyplot` interface and some examples

In our Jupyter Notebook, the first thing that we notice is that we have a command that includes `matplotlib inline` with the % sign as shown in the following code block:

```
%matplotlib inline
```

This is basically the way we tell the Jupyter Notebook that we want to see the plots in the notebook. When we don't use this command and execute this line, we see that the plot will appear in a different window.

Pyplot is basically a collection of command style functions that make `matplotlib` work similar to MATLAB. The idea is that we have a collection of functions and each function makes some changes to a figure, and this figure is considered to be the current figure. So every function does something to a figure; for instance, we can create a figure, we can create a plotting area in a figure, we can plot a line in a subplot of the figure, and we can change the labels. We have to keep in mind which is the current figure when we use the `pyplot` interface. Let's take a look at some examples:

1. The first convention to import `matplotlib` into the current session is `matplotlib.pyplot` as plt:

   ```
   import matplotlib as plt
   ```

2. Our first command includes the `plot` function from the `plt` module and `pyplot` module, and we will also pass a list of numbers. So, when we execute the line, we will see that the command creates a figure. In the following diagram, we have a figure even though we cannot see it, inside the figure we have a subplot, and inside this subplot we have a line plot that is just a graphical representation of the numbers we have in this list:

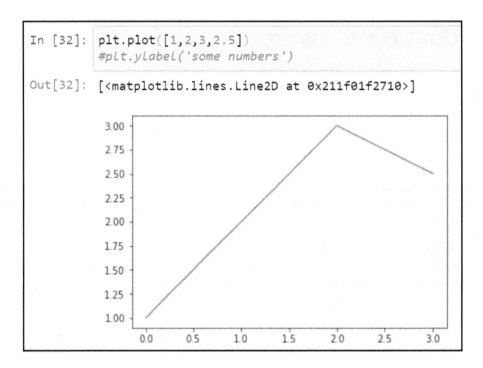

```
In [32]: plt.plot([1,2,3,2.5])
         #plt.ylabel('some numbers')

Out[32]: [<matplotlib.lines.Line2D at 0x211f01f2710>]
```

In the preceding diagram, we can see what this function did, and this figure is considered to be the current figure. Every other function that we use or apply, for instance, if we call this function plt.ylabel, would place a label in the y axis and will be known as ylabel. The label, in this case, is some numbers. Let's run it again and view the following diagram, wherein we can see how this function is applied or has made the label appear in the y axis, as shown in the following diagram:

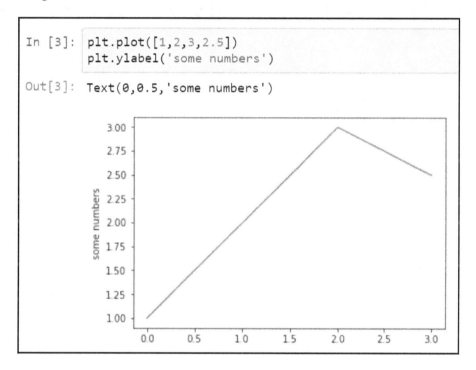

```
In [3]:   plt.plot([1,2,3,2.5])
          plt.ylabel('some numbers')

Out[3]:   Text(0,0.5,'some numbers')
```

3. The most commonly-used function in the `pyplot` interface is the `plot` function, which can take many arguments. For instance, if we pass two lists of numbers, it will assume that the first list is the x coordinates and the second list is the y coordinates, and, in this case, it will plot a line by default. So, let's take a look at the `plot` function in the following diagram:

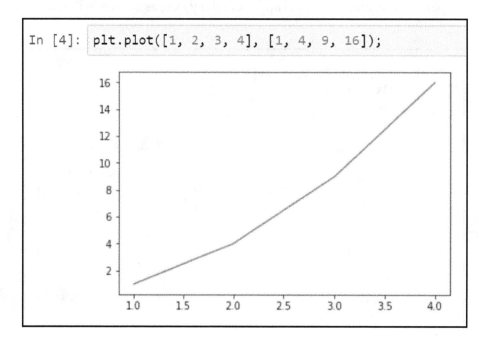

```
In [4]:  plt.plot([1, 2, 3, 4], [1, 4, 9, 16]);
```

This interface thus works by building a graph called a sequence of functions and all of them are applied to the current figure or current subplot. So, let's see how we usually construct a graph using `pyplot` by performing the following steps:

1. We will add two lists to the `plot` function and, as we saw previously, the default behavior is to plot a line; hence we can see the x values and y values in which we are stating that we want the line to be of a `lightblue` color, and we want the `linewidth` to be of size 3. We can view the output in the following diagram:

```
In [34]:  plt.plot([1, 2, 3, 4], [10, 20, 25, 30], color='lightblue', linewidth=3)
          #plt.scatter([0.3, 3.8, 1.2, 2.5], [11, 25, 9, 26], color='darkgreen', marker='^')
          #plt.xlim(0.5, 4.5)
          #plt.title("Title of the plot")
          #plt.xlabel("This is the x-label")
          #plt.ylabel("This is the y-label");

Out[34]:  [<matplotlib.lines.Line2D at 0x211f0277908>]
```

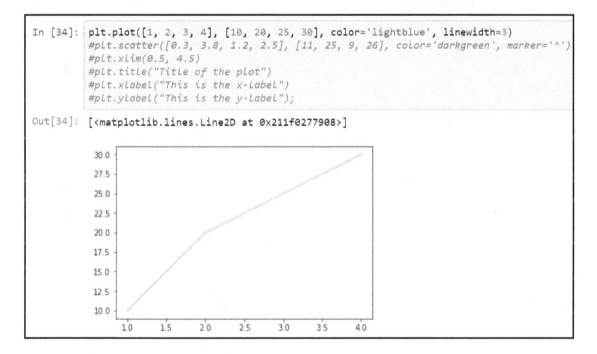

In the preceding diagram, we can see that we have commented on every other function. We will now remove the comments line by line so that we can see all the changes that are applied to every function that we use.

2. The scatter plot plots the coordinates that we provide instead of a line. In this case, they are the X-coordinates and the Y-coordinates. It will also create plot markers. We will then add an argument that will define the markers to be colored in darkgreen. Let's now view the plotting of the scatter plot in the following diagram:

```
In [35]:  plt.plot([1, 2, 3, 4], [10, 20, 25, 30], color='lightblue', linewidth=3)
          plt.scatter([0.3, 3.8, 1.2, 2.5], [11, 25, 9, 26], color='darkgreen', marker='^')
          #plt.xlim(0.5, 4.5)
          #plt.title("Title of the plot")
          #plt.xlabel("This is the x-label")
          #plt.ylabel("This is the y-label");

Out[35]:  <matplotlib.collections.PathCollection at 0x211f0471320>
```

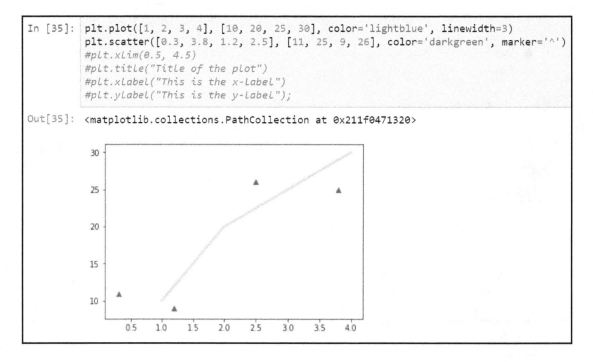

3. We will now apply another function, which will change the x limits of the plot wherein the minimum value is 0.5 and the maximum value is 4.5, so that now the plot will go from 0.5 to 4.5 in the *x* axis. In the following diagram, we will see this modification and also remove the comments line by line:

```
In [35]: plt.plot([1, 2, 3, 4], [10, 20, 25, 30], color='lightblue', linewidth=3)
         plt.scatter([0.3, 3.8, 1.2, 2.5], [11, 25, 9, 26], color='darkgreen', marker='^')
         plt.xlim(0.5, 4.5)
         #plt.title("Title of the plot")
         #plt.xlabel("This is the x-label")
         #plt.ylabel("This is the y-label");
```

```
Out[35]: <matplotlib.collections.PathCollection at 0x211f0471320>
```

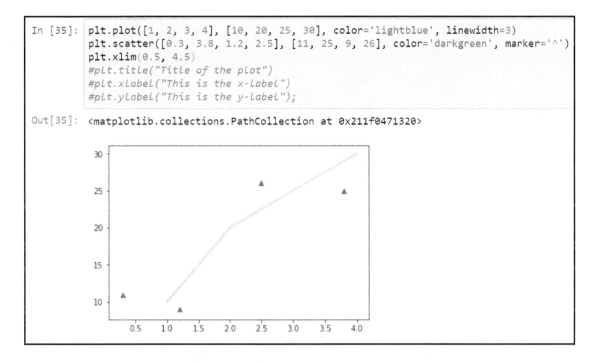

4. The next function will be to add a plot title and to change the plot label. In the following diagram, we can see how we have added the title, the x label, and the y label of the plot:

```
In [36]:  plt.plot([1, 2, 3, 4], [10, 20, 25, 30], color='lightblue', linewidth=3)
          plt.scatter([0.3, 3.8, 1.2, 2.5], [11, 25, 9, 26], color='darkgreen', marker='^')
          plt.xlim(0.5, 4.5)
          plt.title("Title of the plot")
          plt.xlabel("This is the x-label")
          plt.ylabel("This is the y-label");
```

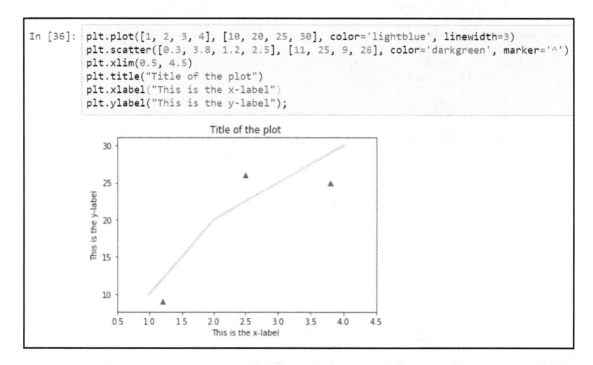

When we work with just one subplot or one figure, it is generally very convenient to use the `pyplot` interface. But when we're working with more than one figure or more than one subplot in a single figure, it can get confusing.

5. We will now run the example and view it in the following screenshot:

```
In [6]:  import numpy as np
         def f(t):
             return np.exp(-t) * np.cos(2*np.pi*t)

         t1 = np.arange(0.0, 5.0, 0.1)
         t2 = np.arange(0.0, 5.0, 0.02)
```

6. Now that we have the data shown in the preceding diagram, we will now produce a figure with two subplots. We will first create a figure and, with this command, we will add a subplot. What we want is a grid of subplots that consists of a grid of two rows and one column, and the plot that we will be using will be the first one. Now, in the first plot, we want to plot t1 and f (t) that we generated and defined, as can be seen in the following screenshot:

```
In [6]:  import numpy as np
         def f(t):
             return np.exp(-t) * np.cos(2*np.pi*t)

         t1 = np.arange(0.0, 5.0, 0.1)
         t2 = np.arange(0.0, 5.0, 0.02)

In [7]:  plt.figure()
         plt.subplot(2, 1, 1)
         plt.plot(t1, f(t1), 'bo')

         plt.subplot(2, 1, 2)
         plt.plot(t2, np.cos(2*np.pi*t2), 'r--')

         plt.ylabel("Y label"); # Which subplot is modifying this function?
```

7. In the preceding diagram, we will use another command to change the argument in the plot. We will still have the grid of two rows and one column, which will be the second plot, we will plot the object t2, and we will also use the marker. In order to change the label, we need to keep the current subplot in mind:

```
In [7]:  plt.figure()
         plt.subplot(2, 1, 1)
         plt.plot(t1, f(t1), 'bo')

         plt.subplot(2, 1, 2)
         plt.plot(t2, np.cos(2*np.pi*t2), 'r--')

         plt.ylabel("Y label"); # Which subplot is modifying this function?
```

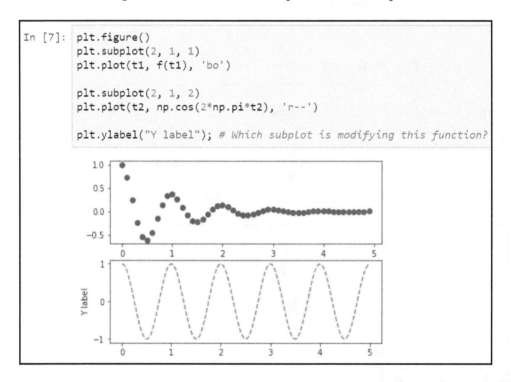

Hence, the object-oriented interface is preferable as we can be sure of which object we are referring to, while in this case, whenever we apply a function, we have to keep in mind the current subplot or the current figure.

Object-oriented interface

We will now look into what object-oriented interface matplotlib is. We will discuss the following topics:

-   The object-oriented interface
-   Creating figures with a grid of subplots
-   Plotting using methods

In the object-oriented interface, what we need to do is create the object and call methods for every object in order to make changes specifically to that object. Let's look at a simple example, where a plot is created using the object-oriented interface. We usually use the `plt.subplots` function to create two objects at the same time. When we create a figure object, we usually call it `fig`, and an axis object is usually called an `ax`. In this case, the default function would be to produce a figure with one axis object, or one subplot. As we create our objects, some changes are involved. In order to make changes to our objects, namely, the plot, the title, or to set the labels, we use methods on these objects. Let's go through a few examples, with the diagrams as our point of reference. In this example, we will use the `plot` method on the object to plot NumPy array's x and y1 array, and we want them in a red line. The next will be while using the same object, and this time we want to plot x with y2, and we want them in a blue line. Then, for the same object we will set an x label and a y label and also the title. Let's look at the following diagram to see the result:

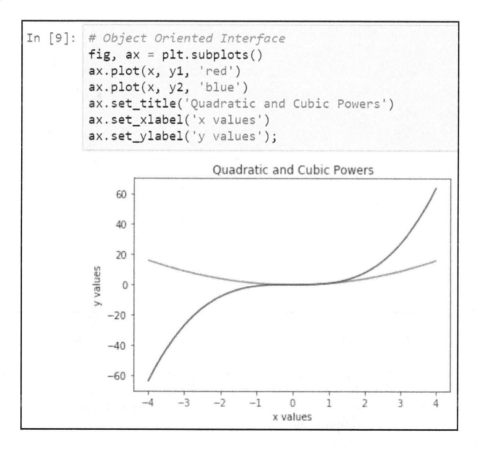

Now, let's see how we can reproduce the same plot using the `pyplot` interface. In the `pyplot` interface, we don't need an explicit command to create a figure because the `plot` function will carry it out. We can also pass many arguments in these functions. First, we will pass the first two pairs of coordinates, the x and `y1` coordinates, and we state that we want this to be a red line. Then we pass the second pair of objects, and we state that we want this to be a blue line. Then we can set the title and the x `label` and the y `label` for this object. Let's view the output in the following diagram:

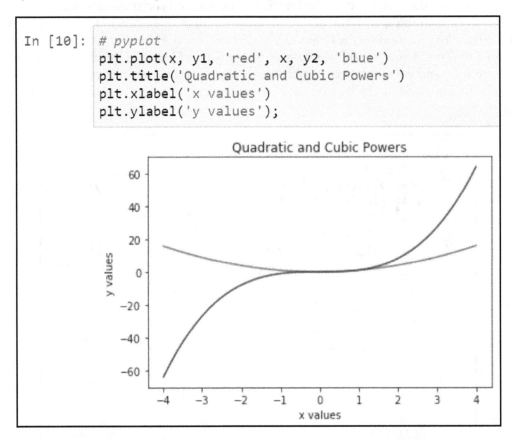

```
In [10]:  # pyplot
          plt.plot(x, y1, 'red', x, y2, 'blue')
          plt.title('Quadratic and Cubic Powers')
          plt.xlabel('x values')
          plt.ylabel('y values');
```

As we reproduce the same plot, we can see that we have fewer lines of code. And the question that arises is the following: *What is the advantage of using the object-oriented interface?*. We come to the conclusion that the usefulness of the object-oriented interface is clearer when we work with multiple subplots in one figure. So, if we're producing just one subplot in one figure, then perhaps it is okay to use the object-oriented, `pyplot` interface. But if we are using many subplots, we will see the advantages of using the object-oriented interface.

So, in order to produce a grid of subplots, we will use the function `plt.subplots`, and can mention how many rows and columns we want, and this will produce a grid of subplots. Hence, in the following diagram, we have four subplots:

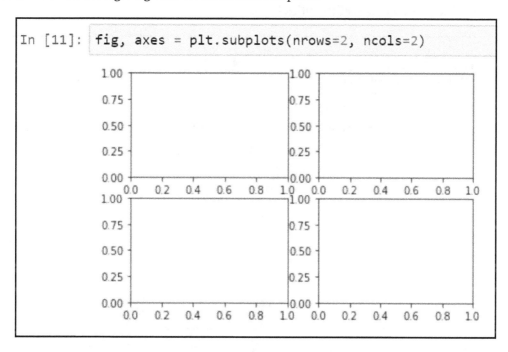

Now, the advantage of using the object-oriented interface is that we can refer to every one of these objects. Let's look at another example where we will run this code again to create two objects, the figure and the axes, and, in this case, the axes will be a multi-dimensional NumPy array with the four subplots. In order to access the subplots, we have to use the NumPy array notation. So, the first subplot is the subplot with the index 0, 0, and, if we want to make some modifications to that object, we need to use the appropriate method. So, in this case, let's say that we want to change the title. We will set the title for this object, that is, in the index 0, 0, and we will set the title to `Upper Left`. We can refer to each of the subplots by its index, and we can use the method that we want in this case. We will change the title for each of these objects. And the nice thing about this is that since the object, `axes`, is a NumPy array, we can iterate whether we want to make similar changes to every object, to every subplot.

So, we can use the flat attribute for every subplot, and we call it `ax`. In this case, we are setting the `xticks` to an empty list, and the `yticks` to an empty list to remove them. When we run this code, as we can see, we don't have the `xticks` or the `yticks` any more, and we have all the titles because we set the titles here:

```
In [12]:   fig, axes = plt.subplots(nrows=2, ncols=2)
           axes[0,0].set_title('Upper Left')
           axes[0,1].set_title('Upper Right')
           axes[1,0].set_title('Lower Left')
           axes[1,1].set_title('Lower Right')

           # To iterate over all items in a multidimensional numpy array, use the `flat` attribute
           for ax in axes.flat:
               # Remove all xticks and yticks...
               ax.set(xticks=[], yticks=[])

           fig.tight_layout();
```

Since we created the figure object, we can use figure methods such as `tight_layout`, which will take care of the way the subplots appear so that they don't overlap. This is basically the idea behind the object-oriented interface; we create the objects and then we apply methods for every object that we create. In the following diagram, we have more examples, so let's see what we produce:

```
In [13]: x = np.linspace(start=-5, stop=5, num=150)
```

```
In [14]: # All functions in one axes
fig, ax = plt.subplots(figsize = (7,4))
ax.plot(x, x, label='Linear')
ax.plot(x, x**2,  label='Quadratic')
ax.plot(x, x**3, label='Cubic')
ax.plot(x, x**4, label='4th Power')
ax.legend();
```

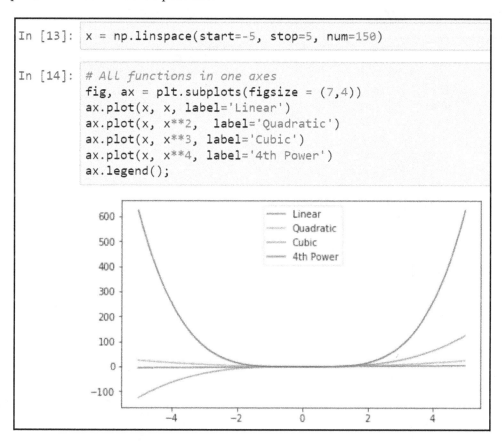

Let's now produce a single figure with a single subplot, in which we will plot all the first four powers of x. We will now see how to plot in different subplots. In the cell, we will create a subplot that has a grid of two rows and two columns, so we will have four subplots in total. Now, we will use the object-oriented interface to set the title for the first one, and the plot for the Linear function. For the second plot, located at the index 0,1 0,1,, we will set the title and plot x and x squared. We can see the result in the following diagram:

In [15]:
```
# 4 Axes/Subplots: One function in one axes
fig, axes = plt.subplots(nrows=2, ncols=2, figsize = (7,4.5))
axes[0,0].set_title('Linear')
axes[0,0].plot(x, x)
axes[0,1].set_title('Quadratic')
axes[0,1].plot(x, x**2)
axes[1,0].set_title('Cubic')
axes[1,0].plot(x, x**3)
axes[1,1].set_title('4th Power')
axes[1,1].plot(x, x**4)
fig.tight_layout();
```

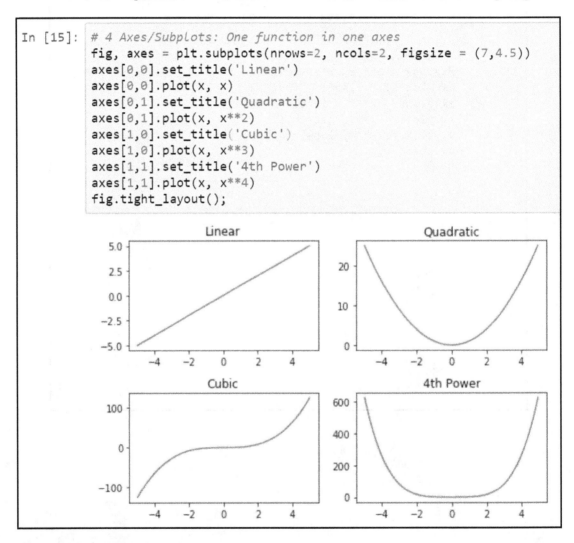

Thus, we can see in the preceding diagram that there are four subplots and one function for every subplot. And we can do more interesting things, which we can view in the following diagram, such as taking a look at the first ten powers of x. So this is essentially how the object-oriented interface works, and this is the way we are going to be using `matplotlib` in the rest of the course:

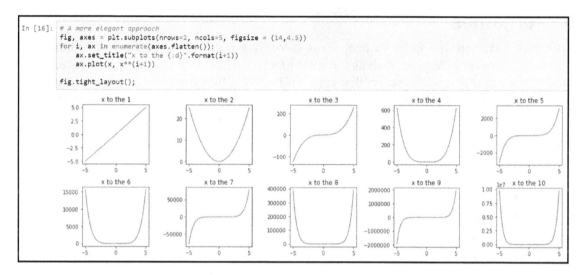

```
In [16]:  # A more elegant approach
          fig, axes = plt.subplots(nrows=2, ncols=5, figsize = (14,4.5))
          for i, ax in enumerate(axes.flatten()):
              ax.set_title("x to the {:d}".format(i+1))
              ax.plot(x, x**(i+1))

          fig.tight_layout();
```

Hence, we have seen the convenience of using the object-oriented interface.

Common customizations

One nice thing about `matplotlib` is that it allows us to tweak every single element of a plot. We will see some of the common customizations that you will always do when working with `matplotlib` when performing data analysis.

First, let's generate some data to work on using the following lines of code:

```
# Generating data
x = np.linspace(-np.pi, np.pi, 200)
sine, cosine = np.sin(x), np.cos(x)
```

We will now look into each customization feature in `matplotlib`.

Colors

Colors are associated with everything that is plotted in the figures. Matplotlib supports a very robust language for specifying colors that should be familiar to a wide variety of users.

Colornames

The most common way whereby people use colors in `matplotlib` is according to initial or name. So, for every element, such as the color of a bar, a line, or a text, you can pass an additional argument with a string that contains the color that you want. If you want your element to be the color blue, you can pass the string `b`, or you can pass the whole word `blue`. The same goes for green; you can pass `g` or the whole word `green`.

The following is a list of some commonly used colors and their initials:

- `b`: blue
- `g`: green
- `r`: red
- `c`: cyan
- `m`: magenta
- `y`: yellow
- `k`: black
- `w`: white

Other colornames that are allowed are the HTML/CSS colornames, such as *burlywood* and *chartreuse*.

 There is a list of 147 such colors; you can see them at `https://www.w3schools.com/tags/ref_colornames.asp`.

Another way you can specify colors in `matplotlib` is using hexadecimal values, such as the ones that you use in HTML or CSS, if you're familiar with those languages. So, you can pass codes such as `#0000FF` and `matplotlib` will understand.

A gray level can be given instead of color by passing a string representation of a number between 0 and 1. 0.0 is black, and 1.0 is white. Every value in-between would be some shade of gray; for example, 0.75 would be a lighter shade of gray.

You can also pass RGB tuples, where the last value will be the transparency value, also known as *alpha*. If you pass a tuple of four elements, matplotlib will interpret the first three elements as an RGB color, and the last one as an alpha value. So, let's say that we want to color the first line, the line for the sine function, as red. We can use this argument, color='red', and then for the cosine function, we use color='#165181':

```
In [20]:  fig, ax = plt.subplots()
          ax.plot(x, sine, color='red')
          ax.plot(x, cosine, color='#165181');
```

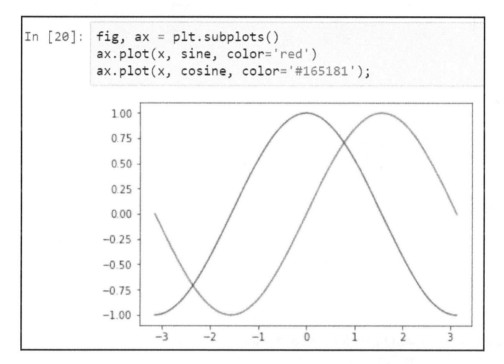

On running the cell, we get the preceding diagram, which shows the line plot.

Setting axis limits

You can change the *x* axis limit with the `set_xlim()` method, and the *y* axis limit with `set_ylim`. These methods take two values, the minimum value, and the maximum value, as shown in the following screenshot:

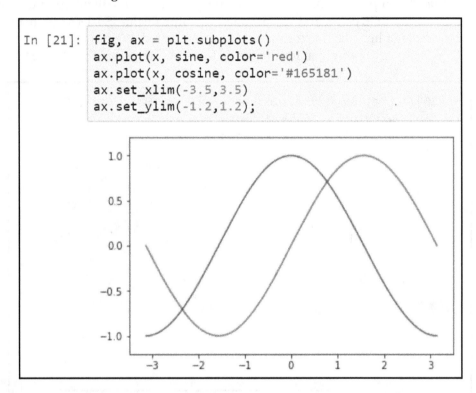

We see that the default limits for both of our axes are changed.

Setting ticks and tick labels

Matplotlib tries to give us very reasonable ticks and tick marks. Ticks are the marks that you can see on the horizontal axis and the tick marks, or the tick labels, are the numbers or the values that correspond to the tick marks. We can also change the tick marks and the tick labels as per requirements:

```
In [22]:  fig, ax = plt.subplots()
          ax.plot(x, sine, color='red')
          ax.plot(x, cosine, color='#165181')

          ax.set_xlim(-3.5,3.5)
          ax.set_ylim(-1.2,1.2)

          ax.set_xticks([-np.pi, -np.pi/2, 0, np.pi/2, np.pi])
          ax.set_yticks(np.arange(-1,1.1,0.5));
```

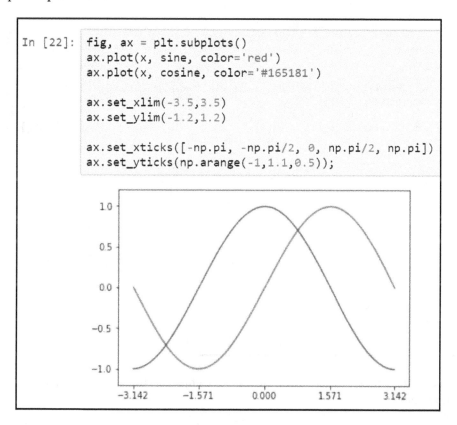

In the preceding diagram, we have our custom set of tick marks and tick labels on both the *x* axis and the *y* axis.

Just as you can change the xticks' positions, you can also change the xticks' labels. Suppose you want the labels on the *x* axis to appear in mathematical notations such as -pi, and -pi/2. You can use the `set_xticks`, `set_yticks`, `set_xticklabels`, and `set_yticklabels` methods to change and assign new values:

```
In [23]:  fig, ax = plt.subplots()
          ax.plot(x, sine, color='red')
          ax.plot(x, cosine, color='#165181')

          ax.set_xlim(-3.5,3.5)
          ax.set_ylim(-1.2,1.2)

          ax.set_xticks([-np.pi, -np.pi/2, 0, np.pi/2, np.pi])
          ax.set_yticks([-1,0,1])

          ax.set_xticklabels([r'$-\pi$', r'$-\pi/2$', r'$0$', r'$+\pi/2$', r'$+\pi$'], size=17)
          ax.set_yticklabels(['-1','0','+1'], size=17);
```

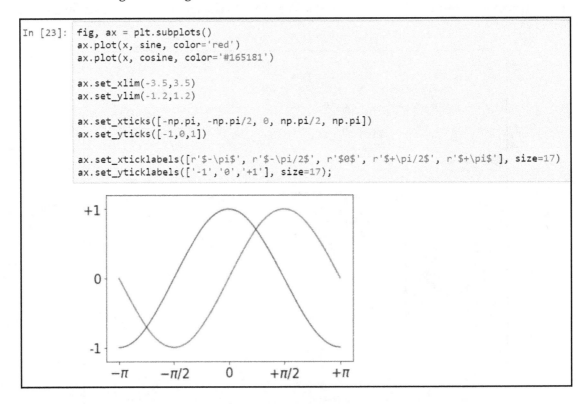

You will see in the preceding output that we have changed the values of tick labels in mathematical notations.

Legend

The legend is used to identify the curves in your plot. It will be a good idea for the viewer of the graph to know that the red curve refers to the sine function and that the blue curve belongs to the cosine function. Hence, we will specify a label for every line that we plot. You can use the `legend()` method with an optional argument where you tell `matplotlib` where you want the label to be placed:

```
In [24]: fig, ax = plt.subplots()
         ax.plot(x, sine, color='red', label='Sine')
         ax.plot(x, cosine, color='#165181', label='Cosine')

         ax.set_xlim(-3.5,3.5)
         ax.set_ylim(-1.2,1.2)

         ax.set_xticks([-np.pi, -np.pi/2, 0, np.pi/2, np.pi])
         ax.set_yticks([-1,0,1])

         ax.set_xticklabels([r'$-\pi$', r'$-\pi/2$', r'$0$', r'$+\pi/2$', r'$+\pi$'], size=17)
         ax.set_yticklabels(['-1','0','+1'], size=17)

         ax.legend(loc='upper left');
```

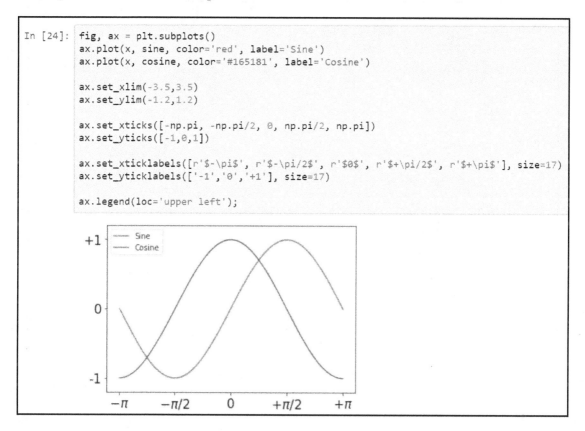

So, in the preceding diagram, we see the label for the sine and cosine functions placed on the upper left of the graph, where we can easily identify each line by the name.

Annotations

There are two main functions or methods for doing annotations in your plot. You can use the text () method by specifying the x coordinate and the y coordinate where you want your text to be shown, and the third argument for this method is the text you actually want to draw in your plot:

```
In [25]: fig, ax = plt.subplots()
         ax.plot(x, sine, color='red', label='Sine')
         ax.plot(x, cosine, color='#165181', label='Cosine')

         ax.set_xlim(-3.5,3.5)
         ax.set_ylim(-1.2,1.2)

         ax.set_xticks([-np.pi, -np.pi/2, 0, np.pi/2, np.pi])
         ax.set_yticks([-1,0,1])

         ax.set_xticklabels([r'$-\pi$', r'$-\pi/2$', r'$0$', r'$+\pi/2$', r'$+\pi$'], size=17)
         ax.set_yticklabels(['-1','0','+1'], size=17)

         ax.legend(loc='upper left')

         ax.text(-0.25,0,'(0,0)') # x coord, y coord,
         ax.text(np.pi-0.25,0, r'$(\pi,0)$', size=15)

         ax.annotate('Origin',
                     xy=(0, 0), # where the arrow points to
                     xytext=(1, -0.7), # location of text
                     arrowprops=dict(facecolor='blue'));
```

In the preceding diagram, the annotation (0,0) is produced using the ax.text(-0.25,0,'(0,0)') method and the annotation (Π,0) is produced using the ax.text(np.pi-0.25, 0, r'$(\pi,0)$', size=15) method. The other function for making annotations in your plots is the annotate method.

The `annotate` method takes many arguments. In the output that was seen previously, we have specified the coordinates where we want the text to be shown using `xytext=(1, -0.7)` and, in `arrowprops=dict(facecolor='blue'))`, we passed a dictionary for the arrow.

Producing grids, horizontal, and vertical lines

There are a couple of methods for producing grids, horizontal, and vertical lines. The `axhline()` method will produce horizontal lines and you can specify other arguments, such as alpha and color, for the color and the transparency. Similarly, we have the `axvline()` method that plots vertical lines along with the arguments:

```
In [26]:  fig, ax = plt.subplots()
          ax.plot(x, sine, color='red', label='Sine')
          ax.plot(x, cosine, color='#165181', label='Cosine')

          ax.set_xlim(-3.5,3.5)
          ax.set_ylim(-1.2,1.2)

          ax.set_xticks([-np.pi, -np.pi/2, 0, np.pi/2, np.pi])
          ax.set_yticks([-1,0,1])

          ax.set_xticklabels([r'$-\pi$', r'$-\pi/2$', r'$0$', r'$+\pi/2$', r'$+\pi$'], size=17)
          ax.set_yticklabels(['-1','0','+1'], size=17)

          ax.legend(loc='upper left')

          ax.text(-0.25,0,'(0,0)') # x coord, y coord,
          ax.text(np.pi-0.2,0.05, r'$(\pi,0)$', size=15)

          ax.annotate('Origin',
                      xy=(0, 0), # where the arrow points to
                      xytext=(1, -0.7), # location of text
                      arrowprops=dict(facecolor='blue'));

          ax.axhline(0, color='black', alpha=0.9) #horizontal line
          ax.axvline(0, color='black', alpha=0.9) #vertical
          ax.grid();
```

The `grid()` method will give you a grid for your plot, in other words, the gray lines that you see in the output that make your plot look better. You can customize almost every element in `matplotlib`, but the ones that we have undertaken are the commonly used changes that you will want to do.

EDA with seaborn and pandas

Exploratory Data Analysis (**EDA**) is an approach to analyzing datasets to summarize their main characteristics, often with visual methods. It is used to understand the data, get some context regarding it, understand the variables and the relationships between them, and formulate hypotheses that could be useful when building predictive models.

Understanding the seaborn library

Seaborn is a library that makes attractive and informative graphics that include statistics in Python. Matplotlib was used to build this library. It is also integrated with Python's data science stack. It also has support for NumPy and `pandas`, as well as SciPy's statistical routines and stats models.

 To know more about seaborn and its features, please visit `https://www.` `datasciencecentral.com/profiles/blogs/opensource-python-` `visualization-libraries`.

The following is the import statement for the seaborn library:

```
# standard import statement for seaborn
import seaborn as sns
```

Performing exploratory data analysis

For performing exploratory data analysis, let's use a dataset that contains information about housing prices and the characteristics of different houses that were sold in a city in the United States. This dataset comes in the supplementary material for this book, and we will load this into the `pandas` DataFrame:

```
In [3]:   housing = pd.read_csv('../data/house_train.csv')

In [4]:   housing.shape

Out[4]:   (1460, 81)

In [5]:   housing.info()

          <class 'pandas.core.frame.DataFrame'>
          RangeIndex: 1460 entries, 0 to 1459
          Data columns (total 81 columns):
          Id              1460 non-null int64
          MSSubClass      1460 non-null int64
          MSZoning        1460 non-null object
          LotFrontage     1201 non-null float64
          LotArea         1460 non-null int64
          Street          1460 non-null object
          Alley           91 non-null object
          LotShape        1460 non-null object
          LandContour     1460 non-null object
          Utilities       1460 non-null object
          LotConfig       1460 non-null object
          LandSlope       1460 non-null object
          Neighborhood    1460 non-null object
          Condition1      1460 non-null object
          Condition2      1460 non-null object
          BldgType        1460 non-null object
          HouseStyle      1460 non-null object
          OverallQual     1460 non-null int64
```

This is how we will be loading the dataset in the `pandas` DataFrame. We see that our DataFrame has 1,460 observations and 81 columns. Each column represents a variable in the DataFrame. Here, we are not going to use all the variables; instead, we will focus on just a few for performing exploratory analysis.

Key objectives when performing data analysis

All data analysis must be guided by some key questions or objectives that guide everything that we do. Before starting any data analysis tasks, you must have a clear goal in mind. As an example, the following objectives will guide our exploration of this dataset:

- Understand the individual variables in the dataset
- Understand how the variables in this dataset relate to the sale price of the house

We are going to do this because when you understand your data and the problem, you will be in a good position to get meaningful results out of your analysis.

Types of variable

In general, there are two possible types of variable:

- Numerical variables
- Categorical variables

By numerical variables, we mean variables for which the values are numbers. And, by categorical variables, we mean variables for which the values are categories:

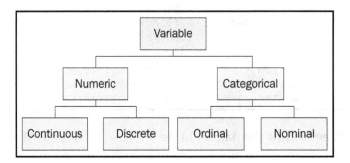

Within these two large categories, we have two sub-categories. For numerical variables, we have continuous variables that can theoretically take any value within an interval. On the other hand, we have discrete numerical variables; these are variables that can take very specific values within an interval. For categorical variables, we again have two types of categorical variables. The first one is what is called the ordinal variable, and these are those variables for which you have a natural order for the categories.

For instance, if you have a variable that is called quality, and the categories for this variable are low quality, medium quality, and high quality, then you know you have an ordinal variable because these categories have a natural order between them. You can say that medium is, in some sense, better than low quality, and that good quality is better than medium and low quality. And finally, nominal categorical variables are those categorical variables that don't have any order between them.

Let's take a look at the following examples of the numerical variables within our dataset:

- SalesPrice: This variable represents the sales price of the house
- LotArea: This variable represents the size of a lot in square feet
- OverallQual: This variable represents rates of the overall material and finish of the house
- OverallCond: This variable represents rates the overall conditions of the house
- 1stFirSF: This variable represents the first floor in square feet
- 2ndFirSF: This variable represents the second floor in square feet
- BedroomAbvGr: This variable represents bedrooms above grade (does not include basement bedrooms)
- YearBuilt: This variable represents the original construction date (this is not technically a numerical variable, but we will use it to produce another variable called Age)

Let's take a look at the following examples of the categorical variables within our dataset:

- MSZoning: This variable identifies the general zoning classification of the sale
- LotShape: This variable represents the general shape of the property
- Neighborhood: This variable represents physical locations within Ames city limits
- CentralAir: This variable represents central air conditioning
- SaleCondition: This variable represents condition of sale
- MoSold: This variable represents month sold (MM)
- YrSold: This variable represents year sold (YYYY)

The reasons why these variables were termed to be numerical and categorical will be seen in the following section of this chapter when we actually analyze the data.

Analyzing variables individually

First, let's define the names of the variables that we are going to use in this analysis. We have a list of the numerical variables and a list of the categorical variables. Then, we will redefine our housing DataFrame with a DataFrame that contains only the variables that we just defined. Then, we use the `shape` attribute to see the size of the new DataFrame:

```
In [7]:  numerical_vars = ['SalePrice','LotArea', 'OverallQual', 'OverallCond',
                            'YearBuilt', '1stFlrSF', '2ndFlrSF', 'BedroomAbvGr']
         categorical_vars = ['MSZoning', 'LotShape', 'Neighborhood', 'CentralAir', 'SaleCondition', 'MoSold', 'YrSold']

In [8]:  housing = housing[numerical_vars+categorical_vars]

In [9]:  housing.shape

Out[9]:  (1460, 15)
```

In the preceding diagram, we can see that the shape of our DataFrame has now changed because we have only 15 columns.

Understanding the main variable

Let's talk about the main variable that we want to understand, the `SalePrice` of the house. The first thing that we do when we have a categorical variable is that we usually want to know their descriptive statistics:

```
In [10]:  #descriptive statistics summary
          housing['SalePrice'].describe()

Out[10]:  count      1460.000000
          mean     180921.195890
          std       79442.502883
          min       34900.000000
          25%      129975.000000
          50%      163000.000000
          75%      214000.000000
          max      755000.000000
          Name: SalePrice, dtype: float64
```

So here, we have an idea of the range of values for this variable. In the preceding diagram, we see that the mean price in the dataset is $180,000 for the average house. We have a standard deviation of around $80,000. The minimum value that corresponds to the cheapest house in the dataset is around $35,000, and the maximum value that corresponds to the most expensive house in the dataset is $755,000.

It is always a good idea to calculate the descriptive statistics of the numerical variables because you will get an understanding of the values this variable can take, as well as the distribution and dispersion of the variable. We must complement the information that we get from the numerical summaries with a graphical representation of our variable. For numerical variables, a typical way to understand the variable visually is with a histogram. To get the histogram of a `pandas` series, we can use the `hist` method, as shown in the following diagram:

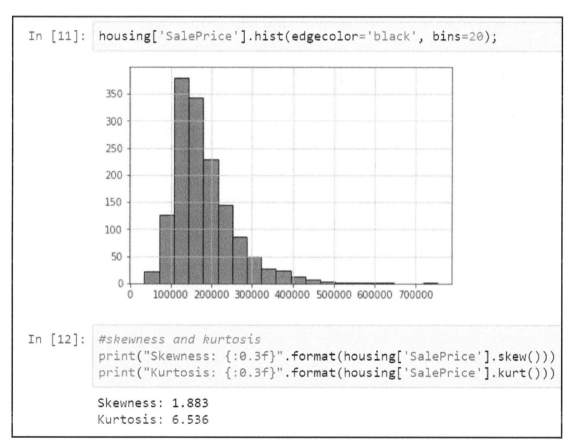

```
In [11]:  housing['SalePrice'].hist(edgecolor='black', bins=20);
```

```
In [12]:  #skewness and kurtosis
          print("Skewness: {:0.3f}".format(housing['SalePrice'].skew()))
          print("Kurtosis: {:0.3f}".format(housing['SalePrice'].kurt()))

          Skewness: 1.883
          Kurtosis: 6.536
```

Here, we see that very few houses are priced below $100,000, so most of the houses are concentrated between $100,000 and $200,000. The other thing that we notice here is that we have very few observations with very high prices. We have few houses above $400,000, and the distribution has a long tail that you can confirm with the following numerical statistics, the Skewness and the Kurtosis. A skewness of 0 will tell you that you have a symmetric distribution, and, if you have a positive value, it will tell you that you have a positive tail similar to the one that we observed here. The kurtosis will tell you the thickness of your distribution, and, as you can see here, we have very high values concentrated around $100,000 and $200,000, which is why we got a kurtosis of 6.5. The usual value for a variable that is close to the normal distribution is around 3.

Numerical variables

If you want to examine all the numerical variables in the dataset, you can do it very easily with the hist method that you get for the pandas DataFrame object. To do so, we are subsetting the housing DataFrame with just a list of the numerical variables that we have, and, just as we did for the pandas series, we can do the same for the pandas DataFrame. We use the describe method, which will show a little DataFrame that contains all the descriptive statistics for every numerical variable that we have in our dataset. Now, if we want to visualize all these variables one by one, one nice thing about pandas and the visualization capabilities that you have with pandas is that you can actually apply the hist method to a DataFrame and not only to a series. So, we use the hist() method with the following arguments: the edgecolor is assigned black, so that the black lines appear between the bars, while the number of bins in the histogram is to be set as 15, the figsize as (14, 5), and the layout as 2 rows by 4 columns, so in total we will have 8 histograms:

```
In [13]: housing[numerical_vars].describe()
```

Out[13]:

	SalePrice	LotArea	OverallQual	OverallCond	YearBuilt	1stFlrSF	2ndFlrSF	BedroomAbvGr
count	1460.000000	1460.000000	1460.000000	1460.000000	1460.000000	1460.000000	1460.000000	1460.000000
mean	180921.195890	10516.828082	6.099315	5.575342	1971.267808	1162.626712	346.992466	2.866438
std	79442.502883	9981.264932	1.382997	1.112799	30.202904	386.587738	436.528436	0.815778
min	34900.000000	1300.000000	1.000000	1.000000	1872.000000	334.000000	0.000000	0.000000
25%	129975.000000	7553.500000	5.000000	5.000000	1954.000000	882.000000	0.000000	2.000000
50%	163000.000000	9478.500000	6.000000	5.000000	1973.000000	1087.000000	0.000000	3.000000
75%	214000.000000	11601.500000	7.000000	6.000000	2000.000000	1391.250000	728.000000	3.000000
max	755000.000000	215245.000000	10.000000	9.000000	2010.000000	4692.000000	2065.000000	8.000000

```
In [14]: housing[numerical_vars].hist(edgecolor='black', bins=15, figsize=(14, 5), layout = (2,4));
```

So, in a single line of code, we can see that `pandas` produces a very nice visualization of the 8 numerical variables that we have in this dataset and, as you can see here, you get a lot of information from watching this, from watching and analyzing this visualization. The following is what we observe for each histogram:

- The distribution of sizes in the first floor, as you can see, is skewed to the right. So what it tells you is that you have very few big houses, but most of the houses have these values around 1,000 or 1,200 square feet.
- You see a big bar around 0 in the `2ndFlrSF` variable. Those are the variables that don't have a second floor.
- Most houses have three bedrooms.
- The lot area is highly skewed; there are few houses with a large amount of area.

- The ratings for conditions and quality tend to be around 5; few houses have very high or low ratings.
- The `YearBuilt` variable is actually not useful in its present form. However, we can use it to construct a variable that actually makes sense; the age of the house at the time of the sale.

So, we will define a new variable, `Age`, as the year in which the house was sold minus the year in which the house was built. Then, we will remove the `YearBuilt` variable from our numerical variables and replace it with the `Age` variable in our list of numerical variables, and do the plot again to see how the distribution of age appears:

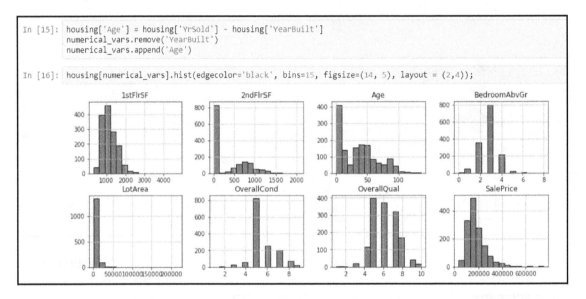

We see the `Age` variable with a big bar here at 0, so this means that many houses were sold; almost 400 houses were sold when they were new.

Categorical variables

The type of plot that is recommended for a categorical variable is the bar plot. To do a bar plot for a categorical variable in `pandas`, you take the `pandas` series object `SaleCondition`, calculate the `value_counts`, and then use the `plot` method;

```
In [17]: housing['SaleCondition'].value_counts().plot(kind='bar', title='SaleCondition');
```

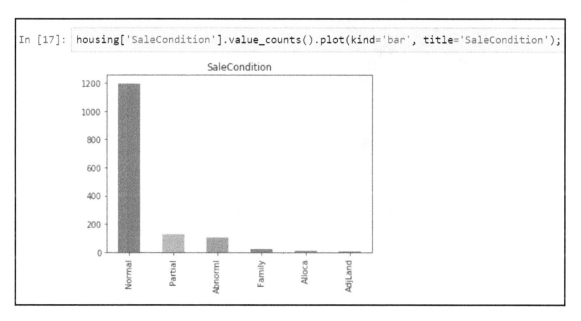

When we execute the line, what we get is a count of the different categories within this `SaleCondition` variable. So, most of the houses were sold under the **Normal** condition, almost 1,200 of them, and very few houses were sold with the other conditions mentioned.

Now, in order to visualize all the categorical variables in our dataset, just as we did with the numerical variables, we don't have a method to do it directly with `pandas`. But, thanks to the things that we learned from our *Introducing Matplotlib* section in this chapter, we can do it using the knowledge that we obtained back there. Hence, we will create a figure and an axis object with the `plt.subplot` function with a grid of 2 rows and 4 columns, which will give us 8 subplots. We then assign the `figsize` as `(14, 6)`. Next, we will write a loop for looping over every categorical variable in every subplot in this `ax.flatten()` object.

So, for every variable, we take the `pandas` series object that will calculate the value count. Then we use the `plot` method with `kind` as `bar`, and the only modification that we have is asking `pandas` to plot the bar plot in the `subplot` object and the `loop` variable. And finally, we have just the `tight_layout` method for the subplots to appear nicely in the diagram:

```
In [18]: fig, ax = plt.subplots(2,4, figsize=(14,6))
         for var, subplot in zip(categorical_vars, ax.flatten()):
             housing[var].value_counts().plot(kind='bar', ax=subplot, title=var)

         fig.tight_layout()
```

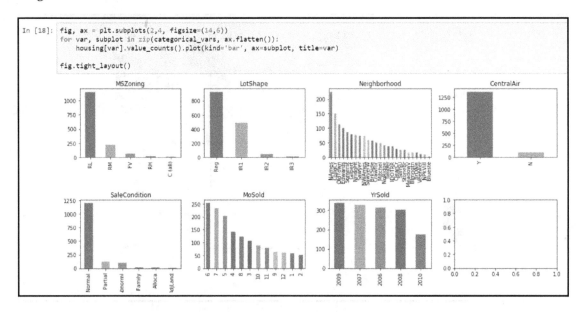

So, as we see here, in just four lines of code, we can produce a very complex and rich visualization.

To have a better understanding between the variables and `SalePrice`, we will write a little function that identifies those levels of category that have less than 30 values. Then we will apply the function to every row in the `pandas` DataFrame, using the `apply` method. Then we have a `for` loop that will detect and keep the observations where the levels have more than 30 observations:

```
In [19]:  def identify_cat_above30(series):
              counts = series.value_counts()
              return list(counts[counts>=30].index)


In [20]:  levels_to_keep = housing[categorical_vars].apply(identify_cat_above30, axis=0)
          levels_to_keep

Out[20]:  MSZoning                                       [RL, RM, FV]
          LotShape                                     [Reg, IR1, IR2]
          Neighborhood       [NAmes, CollgCr, OldTown, Edwards, Somerst, Gi...
          CentralAir                                            [Y, N]
          SaleCondition                      [Normal, Partial, Abnorml]
          MoSold                  [6, 7, 5, 4, 8, 3, 10, 11, 9, 12, 1, 2]
          YrSold                         [2009, 2007, 2006, 2008, 2010]
          dtype: object


In [21]:  for var in categorical_vars:
              housing = housing.loc[housing[var].isin(levels_to_keep[var])]


In [22]:  housing.shape

Out[22]:  (1246, 16)
```

When we execute these lines, we end up with fewer observations; 1,246 observations of 16 variables.

Now, let's look again at the same visualization we saw previously:

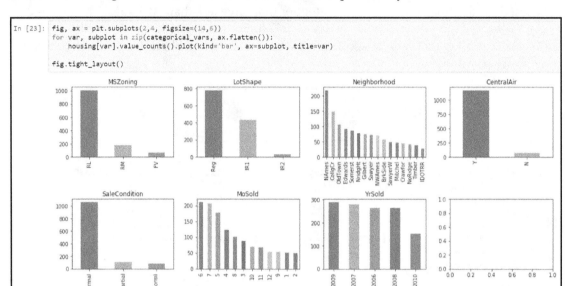

```
In [23]: fig, ax = plt.subplots(2,4, figsize=(14,6))
         for var, subplot in zip(categorical_vars, ax.flatten()):
             housing[var].value_counts().plot(kind='bar', ax=subplot, title=var)

         fig.tight_layout()
```

As we can now see, we don't have those levels where we had fewer than 30 observations. Now, the `MSZoning` variable has only three levels, the `LotShape` variable has only three levels, and the `SaleCondition` variable has only three levels as well.

Relationships between variables

The relationships between different variables can be visualized using the namespace plot from `matplotlib`. The scatter plot is used for visualizing relationships between two numerical variables, and the box plot is used for visualizing relationships between one numerical variable and one categorical variable. The complex conditional plot will be used to visualize many variables in a single visualization.

Scatter plot

To produce a scatter plot with `pandas`, all you have to do is to use the plot namespace. Within the plot namespace, you have a `scatter()` method and pass an x value and a y value:

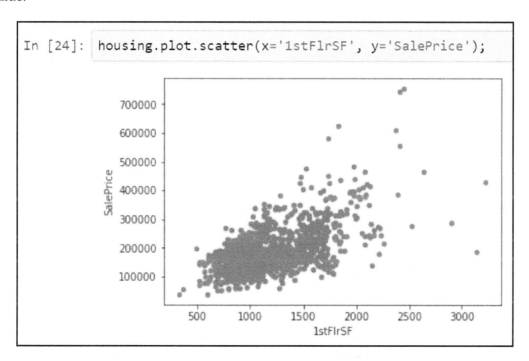

```
In [24]:   housing.plot.scatter(x='1stFlrSF', y='SalePrice');
```

We see here that we have a positive relationship between the `1stFlrSF` of the house and the `SalePrice` of the house. So, the `1stFlrSF` variable is plotted on the *x* axis, the `SalePrice` variable on the *y* axis, and we can see here clearly that there is a positive relationship between these two variables; the more `1stFlrSF` you have, the higher the likely sale price of the house.

Seaborn also provides us with a nice function called `jointplot`, which will give you a scatter plot, but, in addition, will also give you a marginal plot:

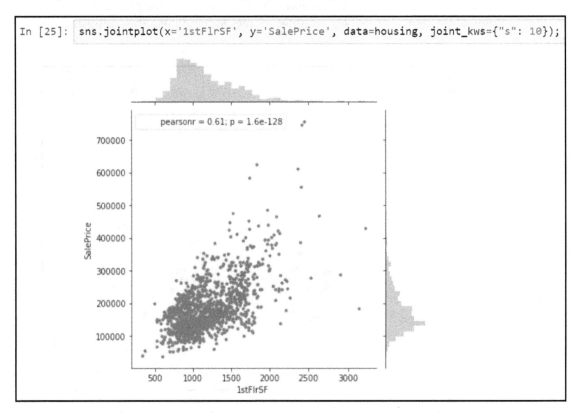

```
In [25]:   sns.jointplot(x='1stFlrSF', y='SalePrice', data=housing, joint_kws={"s": 10});
```

In the margins, you will see the distribution of your variables in the *x* axis and the *y* axis. This is nice because you not only see the relationships between the two variables, but you can also see how they are distributed individually.

If you want to visualize many scatter plots at the same time, you can produce what is called a scatter plot matrix, and seaborn makes this very easy to do with the `pairplot` function. You can pass a DataFrame with some numerical variables, and what you will get is a scatter plot matrix:

```
In [26]:   sns.pairplot(housing[numerical_vars[:4]], plot_kws={"s": 10});
```

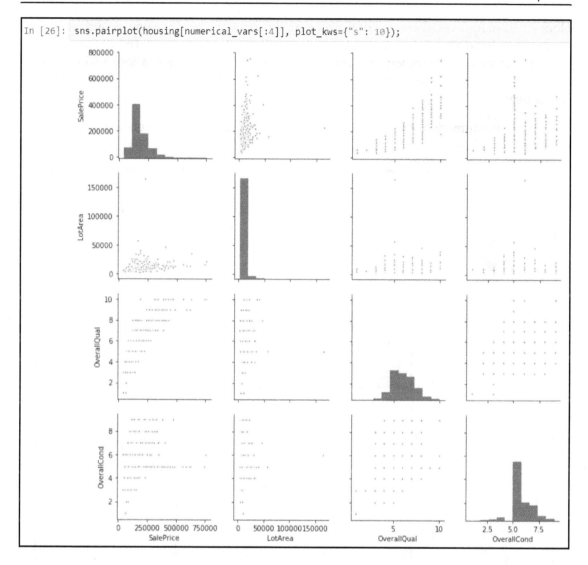

Each of the plots is a pairwise relationship between the variables that you have in your DataFrame.

It is recommended that you use no more than four or five variables at the same time because otherwise it will be very hard to see.

Here we have four variables, one of them being the `SalePrice` of the house, and, as we can see here, we have a very clear positive relationship between the `SalePrice` of the house and another variable. Next, we have the `LotArea`, which shows the relationship between the `OverallQual` and the `SalePrice` of the house. The relationship is still positive, but is not as clear with the `OverallCond` of the house.

We have more numerical variables, so let's visualize those and see the relationship between them and the `SalePrice`:

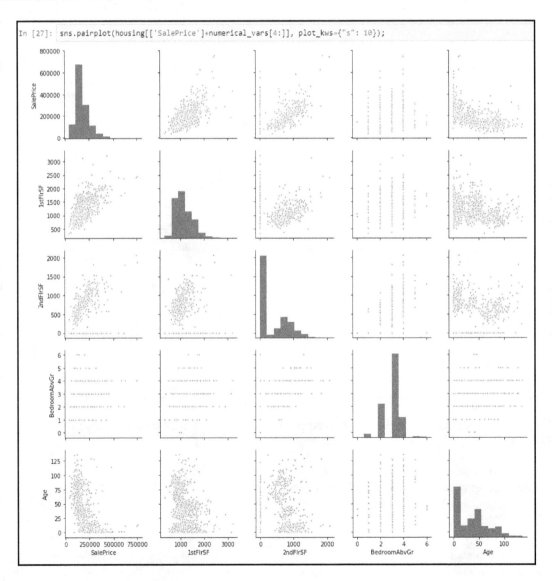

```
In [27]:   sns.pairplot(housing[['SalePrice']+numerical_vars[4:]], plot_kws={"s": 10});
```

Here, we see that the `SalePrice` of the house has a very clear positive relationship with these two variables; the `1stFlrSF` of the house and the `2ndFlrSF`. We also have a negative relationship that is not linear between the `Age` and `SalePrice` variables. We can see here that as the `Age` goes up, the `SalePrice` of the house goes down, but this relationship is a curve, not a straight line.

Box plot

Here, we are interested in the relationship between the categorical variables in our dataset and the `SalePrice` of the house. The standard plot to examine the relationship between a numerical and a categorical variable is the box plot. A box plot is a convenient way of graphically depicting groups of numerical data through their quartiles. Box plots may also have lines extending vertically from the boxes (whiskers), indicating variability outside the upper and lower quartiles, hence the terms box-and-whisker plot and box-and-whisker diagram. Outliers may be plotted as individual points. Box plots are non-parametric; they display variation in samples of a statistical population without making any assumptions as to the underlying statistical distribution.

Now, let's visualize the relationship between the `SalePrice` of the house and the rest of the categorical variables that we have using the box plot:

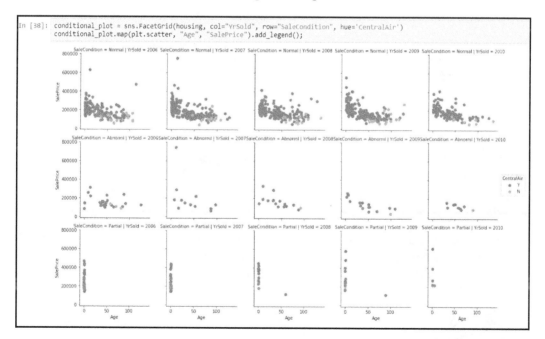

Here, we see the box plot of the variable `CentralAir` that we placed on the *x* axis and the `SalePrice` of the house on the *y* axis. We see that, for houses with no `CentralAir`, the `SalePrice` is definitely lower and the distribution of prices is definitely lower than for houses that have `CentralAir`.

We can also visualize many box plots in a single visualization, as we did previously with scatter plots. Following the same technique of producing a diagram with many subplots, we can iterate through every subplot and ask Python to produce the visualization between the categorical variable and the `SalePrice`:

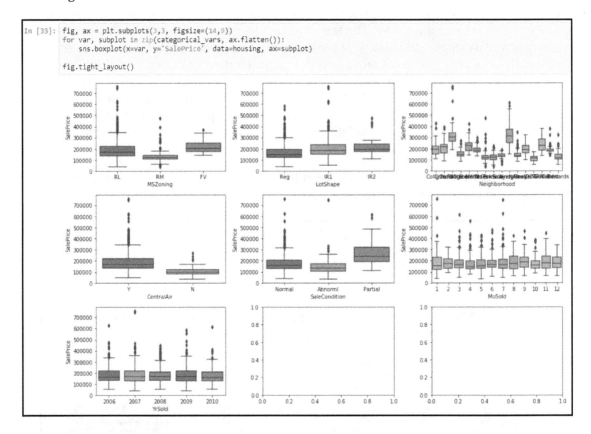

Here, we see a difference in distribution in the category for the variable `MSZoning`;the category `RM` definitely has a distribution with lower sale prices, and, in the `Neighborhood` variable, we see that there is definitely a different distribution for different neighborhoods.

Let's see the relationship between the `Neighborhood` and the `SalePrice` variables:

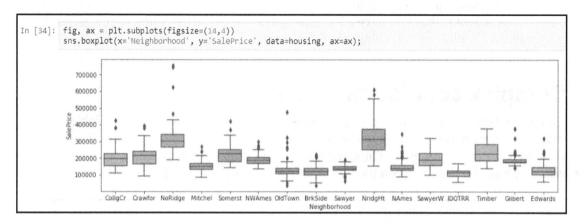

We see that different neighborhoods have varying distributions of price, and it might be a good idea to visualize this in a sorted way, from the cheapest neighborhood to the most expensive neighborhood. This can easily be accomplished using this additional argument `order=sorted`, and we got the order from the housing values sorting the `Neighborhood` by median price:

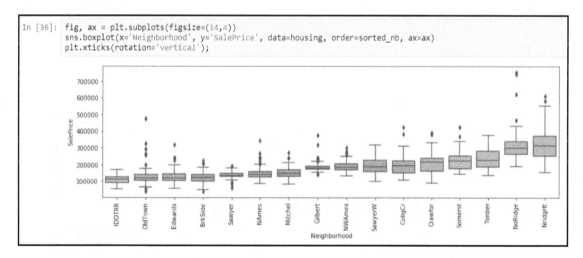

Now, after sorting the neighborhoods according to median price, we see the visualization of the distribution of the prices in the different neighborhoods. Hence, we see the cheapest neighborhoods, where the median price of the house is around $100,000, and the most expensive neighborhoods, where the median price of the house is around $300,000.

We see that for some neighborhoods, dispersion between the prices is very low, so when you see small boxes, this means that all the prices are close to each other, and when you see large boxes, it means that you a lot of dispersion in the distribution of prices. So there is a lot of information we can obtain from these visualizations.

Complex conditional plots

Conditional plots are more complex plots where you can condition on one variable. For example, if we are conditioning in the `Neighborhood`, and you produce conditional plots with the `FacetGrid` function. Thus, we will condition the `Neighborhood`, and then we visualize a scatter plot between the `OverallQual` and the `SalePrice` variables:

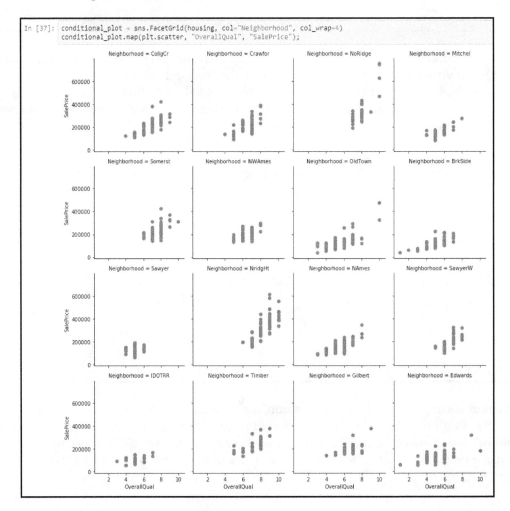

So here, every scatter plot in this visualization is conditioned on each neighborhood. If the positive relationship between two variables holds for the entire neighborhood, then we can say that the observed relationship holds for every neighborhood.

If you want to visualize more variables, you can produce a conditional plot using the `FacetGrid`method and pass categorical variables as columns and rows, with an additional categorical variable to color the points of your scatter plot with different colors. Using the new feature, we will visualize the relationship between `Age` and `SalePrice`, but this time conditioning on the year when the house was sold and the `SaleCondition` variable:

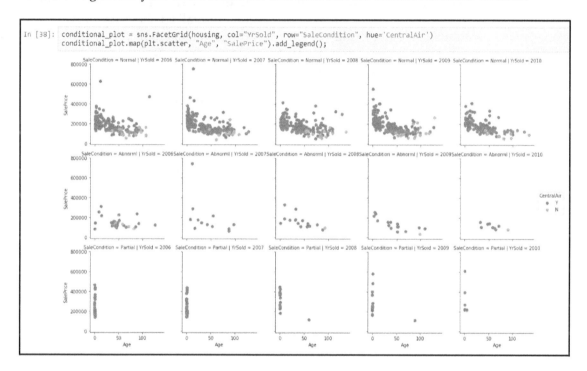

Here, you can see that each column corresponds to each year starting from 2006-2010 and the three rows correspond to the **Normal** condition, the **Abnormal** condition, and the **Partial** condition. We also see that the negative relationship between `Age` and `SalePrice` holds for every subgroup in the normal row and it tends to go down as `Age` increases. The green point corresponds to those houses that do not have `CentralAir`, and since most of the green points are located around 50 to 100, we can say that these are the older houses. One other aspect we can glean from this plot is that we have few observations where the `SaleCondition` was **Partial**, and, for most of these observations, the age of the house was 0.

This is just a sample to show how easy it is to produce complex visualizations and how you can get valuable information from them. You should always try not only to produce these visualizations but to actually get some insight from them as a data analyst.

Summary

In this chapter, we talked about `matplotlib` and the way it works. We also learned about the difference between the object-oriented interface and the `pyplot` interface. In addition, we learned how to produce some common visualizations and talked about how to do visualizations with seaborn and pandas, which are very convenient. Finally, we learned how to visualize the relationships between variables and did a practical example of an exploratory data analysis on a real-world dataset.

In the next chapter, we will look at how to carry out statistical computing with Python.

5
Statistical Computing with Python

In this chapter, we will briefly talk about **Scientific Library for Python (SciPy)**, which is the scientific toolbox for Python. We will get a brief overview of the statistics subpackage and we will use it to perform many statistical calculations, including calculations of probabilities, probability distributions, and confidence intervals. We will also perform hypothesis testing on a real-world dataset, and we will be able to state conclusions that go beyond the sample that we have.

In this chapter, we will cover the following topics:

- SciPy
- Statistics
- Probabilities
- Hypothesis testing
- Performing statistical tests

Introduction to SciPy

SciPy is a tool for doing scientific computing in Python. It is a Python-based ecosystem that is an open source software for math, science, and engineering. It contains various toolboxes dedicated to common issues in scientific computing. So, if you work in any scientific or engineering field, you will likely find the tools you need for doing scientific computing within the subpackages of SciPy. The following are the subpackages of SciPy:

- `scipy.io`: This package provides a tool for dealing with file input/output
- `scipy.special`: This package provides a tool for dealing with special functions
- `scipy.linalg`: This package provides a tool for dealing with linear algebra operations

- `scipy.fftpack`: This package provides a tool for dealing with fast Fourier transforms
- `scipy.stats`: This package provides a tool for dealing with statistics and random numbers
- `scipy.interpolate`: This package provides a tool for dealing with interpolation
- `scipy.integrate`: This package provides a tool for dealing with numerical integration
- `scipy.signal`: This package provides a tool for dealing with signal processing
- `scipy.ndimage`: This package provides a tool for dealing with image processing

Statistics subpackage

We can load or use the statistics subpackage by importing the module known as `stats`. It is a subpackage that contains a large number of probability distributions, as well as a growing library of statistical functions. This is a great tool to use for data analysis or data science with Python because statistics is the core of data science. As we progress in the chapter, we will also learn how to perform some common statistical computations with Python, and we will use this statistics subpackage to make sense of a dataset that contains information about the alcohol consumption of students in Portugal.

We will perform some common statistical calculations using the `stats` subpackage from SciPy in the following order:

1. Introducing the dataset
2. Calculating confidence intervals
3. Performing probability calculations

Lets now look at the Jupyter Notebook and discuss the project, which is about the alcohol consumption of students. In this project, we will use a dataset containing information about particular students from two public schools. This is a real-world dataset that was collected in order to study alcohol consumption in young people and the effects it has on their academic performance. The dataset was built from two sources from school reports and from questionnaires that the students answered.

This dataset has more than 30 variables, and it will help you with your analytical skills. In the following screenshot, we have all our variables in this dataset:

- `school`: This variable represents student's school (binary: Gabriel Pereira `'GP'` or Mousinho da Silveira `'MS'`)
- `sex`: This variable represents student's sex (binary: `'F'` as female or `'M'` as male)
- `age`: This variable represents student's age (numeric: from 15 to 22)
- `address`: This variable represents student's home address type (binary: `'U'` as urban or `'R'` as rural)
- `famsize`: This variable represents family size (binary: `'LE3'` as less or equal to 3 or `'GT3'` as greater than 3)
- `Pstatus`: This variable represents parent's cohabitation status (binary: `'T'` as living together or `'A'` as apart)
- `Medu`: This variable represents mother's education (numeric: 0 as none, 1 as primary education (4th grade), 2—5th to 9th grade, 3—secondary education or 4—higher education)
- `Fedu`: This variable represents father's education (numeric: 0 as none, 1 as primary education (4th grade), 2—5th to 9th grade, 3 as secondary education, or 4 as higher education)
- `Mjob`: mother's job (nominal: `'teacher'`, `'health'` care related, civil `'services'` (for example, administrative or police), `'at_home'` or `'other'`)
- `Fjob`: This variable represents father's job (nominal: `'teacher'`, `'health'` care related, civil `'services'` (for example, administrative or police), `'at_home'` or `'other'`)
- `reason`: This variable represents the reason to choose this school (nominal: close to `'home'`, school `'reputation'`, `'course'` preference, or `'other'`)
- `guardian`: This variable represents a student's guardian (nominal: `'mother'`, `'father'`, or `'other'`)
- `traveltime`: This variable represents home to school travel time (numeric: 1 < 15 min, 2 and 15 to 30 min, 3 as 30 min to 1 hour, or 4—>1 hour)
- `studytime`: This variable represents weekly study time (numeric: 1—< 2 hours, 2—2 to 5 hours, 3—5 to 10 hours, or 4—>10 hours)
- `failures`: This variable represents number of past class failures (numeric: n if 1<=n<3, else 4)
- `schoolsup`: This variable represents extra educational support (binary: yes or no)
- `famsup`: This variable represents family educational support (binary: yes or no)

- `paid`: This variable represents extra paid classes within the course subject (`Math` or `Portuguese`) (binary: yes or no)
- `activities`: This variable represents extra-curricular activities (binary: yes or no)
- `nursery`: This variable represents attended nursery school (binary: yes or no)
- `higher`: This variable represents wants to take higher education (binary: yes or no)
- `internet`: This variable represents internet access at home (binary: yes or no)
- `romantic`: This variable represents a romantic relationship (binary: yes or no)
- `famrel`: This variable represents a quality of family relationships (numeric: from 1—very bad to 5—excellent)
- `freetime`: This variable represents free time after school (numeric: from 1—very low to 5—very high)
- `goout`: This variable represents going out with friends (numeric: from 1—very low to 5—very high)
- `Dalc`: This variable represents workday alcohol consumption (numeric: from 1—very low to 5—very high)
- `Walc`: This variable represents weekend alcohol consumption (numeric: from 1—very low to 5—very high)
- `health`: This variable represents current health status (numeric: from 1—very bad to 5—very good)
- `absences`: This variable represents number of school absences (numeric: from 0 to 93)

These grades are related to the course subject as follows:

- `G1`: This grade represents first-period grade (numeric: from 0 to 20)
- `G2`: This grade represents second-period grade (numeric: from 0 to 20)
- `G3`: This grade represents final-period grade (numeric: from 0 to 20, output target)

However, in this section, we are interested in the following three variables:

- `acl`: We will create this variable from the data for storing the value of alcohol consumption levels
- `G3`: This variable is created for storing the final grade for the course subject
- `gender`: This variable is created for storing the gender of the student

So, let's load the dataset, as shown in the following screenshot:

```
In [5]:  student.head()
```
Out[5]:

	school	sex	age	address	famsize	Pstatus	Medu	Fedu	Mjob	Fjob	...	famrel	freetime	goout	Dalc	Wal
0	GP	F	18	U	GT3	A	4	4	at_home	teacher	...	4	3	4	1	
1	GP	F	17	U	GT3	T	1	1	at_home	other	...	5	3	3	1	
2	GP	F	15	U	LE3	T	1	1	at_home	other	...	4	3	2	2	
3	GP	F	15	U	GT3	T	4	2	health	services	...	3	2	2	1	
4	GP	F	16	U	GT3	T	3	3	other	other	...	4	3	2	1	

5 rows × 33 columns

We will first create the alcohol consumption level variable, in which we will create an intermediate variable called `alcohol_index`, which is just a weighted average of how much the students drink during weekdays and weekends, as shown in the following code block:

```
student.rename(columns={'sex':'gender'}, inplace=True)
student['alcohol_index'] = (5*student['Dalc'] + 2*student['Walc'])/7
# Alcohol consumption level
student['acl'] = student['alcohol_index'] < = 2
student['acl'] = student['acl'].map({True: 'Low', False: 'High'})
```

We will then use this `alcohol_index` variable to divide the students into groups. If the alcohol index is less than or equal to 2, we will call this group the low alcohol consumption group, and if they have an alcohol index greater than 2, we will call this group the high alcohol consumption group.

Confidence intervals

We will start by calculating the confidence intervals. The first variable for which we are going to calculate the confidence interval is the final grade and the statistics that we are interested in, which are the mean. Now let's take a look at `sample_size` in this dataset. We have `649` observations, and because the sample size is greater than 30, we can use the central limit theorem to calculate confidence intervals, as shown in the following screenshot:

```
In [7]:  sample_size = student.shape[0]
         print(sample_size)

         649
```

The statistical theory tells us that we can calculate confidence intervals for the mean of a variable using the normal distribution. To get a confidence interval, we just need the following three numbers:

- The sample mean of the variable that we are interested in
- The standard error for the formula; the standard, and the sample standard deviation divided by the square root of the sample size
- The confidence level

Since we already know how to calculate the sample mean, let's view it in the following screenshot. We can see that the variable for the final grade is `G3` and its output is `11.906009244992296`:

```
In [8]:  sample_mean_grade = student['G3'].mean()
         sample_mean_grade

Out[8]:  11.906009244992296
```

In the following screenshot, we can see just the application of this formula and we will also calculate the standard error as well. So, now we have the first two numbers that we need. To calculate the confidence interval, we will use the `norm` object from the `stats` subpackage. The `norm` object has an `interval()` method that receives three inputs and they are the three numbers that we need:

```
In [13]: std_error_grades = student['G3'].std()/sqrt(sample_size)

In [14]: stats.norm.interval(0.95, loc=sample_mean_grade, scale=std_error_grades)
Out[14]: (11.65745768566587, 12.154560804318722)
```

The first one is the confidence level that we want for our interval which is 0.95. It is usually the number that people use but we can use another number if necessary. The loc parameter will be sample_mean of the variable, and the scale parameter will be std_error that we just calculated. We can see that the 95% confidence interval for the mean of the final grades is from 11.65745768566587 to 12.154560804318722. Now that we know how to calculate a confidence interval for a mean, standard error, and norm, let's look at an example and see how to calculate a confidence interval for a proportion.

We will use the variable, which we just created that is the alcohol consumption level, and look at what the confidence interval will be for the proportion of students with a high alcohol consumption level. The calculation is very similar; we need the following three numbers:

- Sample proportion
- Sample standard deviation
- Confidence level

In this case, for the proportion, the standard error is given by the following formula:

$$SE = \sqrt{\frac{\hat{p}(1 - \hat{p})}{n}}$$

In the preceding formula, p-hat is the sample proportion. Now, let's look at the sample proportion of students with a high alcohol consumption. It is around 0.255778, as you can see in the following screenshot:

```
In [11]: student['acl'].value_counts(normalize=True)

Out[11]: Low      0.744222
         High     0.255778
         Name: acl, dtype: float64
```

We will call the `high_prop` variable and, in the following line, we can use the formula to calculate the standard error and the confidence interval:

```
high_prop = student['acl'].value_counts(normalize=True)['High']
std_error_prop = sqrt(high_prop*(1-high_prop)/sample_size)
```

But this time, we will calculate a 98% confidence interval for the proportion of students in the population that have a high alcohol consumption level:

```
In [13]:  stats.norm.interval(0.98, loc=high_prop, scale=std_error_prop)

Out[13]:  (0.21593666225148048, 0.2956195781183193)
```

The number is from `0.21593666225148048` to `0.29561957811831929`. So 0.25 seems like a good guess for the proportion of students with a high alcohol consumption level in the population.

We must remember that these are calculations in the sample and that we use inferential statistics to say something about the population.

Probability calculations

Now, let's assume that the number in the population is 0.25, and let's use this number to do some probability calculations.

There are dozens of probability distributions available in the `stats` package, and we can use them for doing simulations, to calculate the random variables, and to do probability calculations. Let's now take a look at an example and answer a few questions.

Assuming that the probability of finding a student with a high alcohol consumption level is `0.25`, if we have a class of 10 students, what is the probability of finding five students with high alcohol consumption?

We know we can calculate this probability using the **binomial distribution**. Also in the `stats` package, we have the `binom` object, which has a method called **probability mass function**, or `pmf`, and we provide the parameters that `pmf` needs for the binomial distribution. We are interested in the probability of finding five students that have a high alcohol consumption level in a group of 10 students, with the probability of the population of `0.25`. We will now do the calculation and see the probability in the following screenshot:

```
In [14]:  stats.binom.pmf(k=5, n=10, p=0.25)

Out[14]:  0.058399200439453194
```

We see that the probability of this event is 0.058399200439453194. And, of course, we can calculate many probabilities at the same time. In the following screenshot, we see what the probability of finding the specific number of students is in the x axis, and the height of the bar is the probability of finding this number of students with a high alcohol consumption level in a class of 10. And we can also see that the most likely numbers are found between two and three students in a class of 10. On the other hand, it is very unlikely to find a class full of students with a high alcohol consumption level. We can also see the **Cumulative distribution function** chart adjacent to the **Probability mass function** chart:

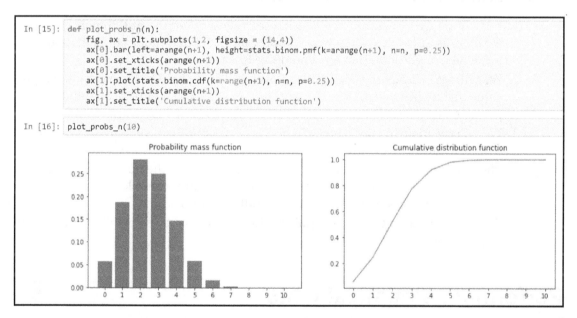

Thus, we can find the cumulative distribution function, and we can use the `cdf()` method from the `binom` distribution to calculate the cumulative distribution function.

Hypothesis testing

In this section, we will perform hypothesis testing to answer this question: "does alcohol consumption affect academic performance?".

We will cover the following topics:

- A null hypothesis testing framework
- Performing a test for the equality of population variances
- Performing a t-test for the equality of population means

We will be using these results to answer the question that we proposed. We will use the Jupyter Notebook and conduct a brief review of the following steps, which we usually follow when we use hypothesis testing to answer a question:

1. We will set up two competing hypotheses; one is called the **null hypothesis** and the other is called the **alternative hypothesis**. What determines these hypotheses is the type of question that we would like to answer.
2. We will set in advance a significance level, called **alpha**. The most common alpha value that is chosen by researchers or data analysts is 5% or 0.05, but other popular choices are 0.01 or 0.1, which is equivalent to 10%.
3. We will calculate the test statistic in the p-value, and then compare this p-value with the value of the alpha.
4. Then, we will move to where we make a decision about the null hypothesis. Based on our comparison of the p-value with the significance level, we decide whether we want to reject or accept the null hypothesis.
5. Finally, we need to state an overall conclusion on what these results imply for the original question.

Performing statistical tests

Before performing a statistical test, let's look at some of the statistical functions that we can use to perform tests using the `scipy` package listed as follows:

- `kurtosistest(a[, axis, nan_policy])`: This package is used to test whether a dataset has normal kurtosis
- `normaltest(a[, axis, nan_policy])`: This package is used to test whether a sample differs from a normal distribution

- skewtest(a[, axis, nan_policy]): This package is used to tests whether the skew is different from the normal distribution
- pearsonr(x, y): This package is used to calculates a Pearson correlation coefficient and the p-value for testing non-correlation
- ttest_1samp(a, popmean[, axis, nan_policy]): This package is used to calculates the t-test for the mean of one group of scores
- ttest_1samp(a, popmean[, axis, nan_policy]): This package is used to calculates the t-test for the mean of one group of scores
- ttest_ind(a, b[, axis, equal_var, nan_policy]): This package is used to calculate the t-test for the means of two independent samples of scores
- ttest_ind_from_stats(mean1, std1, nobs1, ...): This package is used to t-test for means of two independent samples from descriptive statistics
- ttest_rel(a, b[, axis, nan_policy]): This package is used to calculate the t-test on two related samples of scores, a and b
- kstest(rvs, cdf[, args, N, alternative, mode]): This package is used to perform the Kolmogorov-Smirnov test for goodness of fit.
- chisquare(f_obs[, f_exp, ddof, axis]): This package is used to calculate a one-way chi-square test.
- ansari(x, y): This package is used to perform the Ansari-Bradley test for equal scale parameters
- bartlett(*args): This package is used to perform Bartlett's test for equal variances
- levene(*args*, *kwds): This package is used to perform Levene test for equal variances
- shapiro(x[, a, reta]): This package is used to perform the Shapiro-Wilk test for normality
- anderson(x[, dist]): This package is used to perform the Anderson-Darling test for data coming from a particular distribution
- anderson_ksamp(samples[, midrank]): This package is used to perform the Anderson-Darling test for k-samples

These are some of the most popular statistical tests that we can perform, and if we go to the documentation (`https://docs.scipy.org`) of the `scipy.stats` subpackage, we will find many more specialized tests that we can do.

We will use one of these tests to answer the following question: Are the population variances equal in the two groups of students?. Remember that we have two groups of students: one is the group with a low alcohol consumption level, and the other is the group with a high alcohol consumption. Since this is a question about variances, we will use **Bartlett's** test, whose null hypothesis states that the variances in the population are equal:

```
In [18]:  grades_low_acl = student['G3'][student['acl']=='Low']
          grades_high_acl = student['G3'][student['acl']=='High']
          stats.bartlett(grades_low_acl, grades_high_acl)

Out[18]:  BartlettResult(statistic=1.1025085913378183, pvalue=0.29371623181175127)
```

In the preceding screenshot, we see the variances in the sample. They look like they are close to each other, but let's perform a formal test to conclude something about the population variances. As we can see in the screenshot, we have a vector, an object with the grades for students in the low group, a vector with the grades for the students in the high group, and Bartlett's function. The only thing that we have to pass to these functions are the two vectors that we just calculated. Now let's see the result of this test.

We are especially interested in `pvalue`. We see that `pvalue` is `0.2937162318117544`, which is greater than the significance level that we saw before: 5%, or 0.05. So, according to this test, we cannot reject the null hypothesis of equal variances. So, we can assume that the two groups of grades come from a population with the same variances.

In this next test, we will try to answer the following question: Does alcohol consumption affect academic performance?. When we see a visualization of the data or the sample data that we have, we see that there appears to be a very clear difference between the grades, or the mean grades, in the two groups. In the following screenshot, we can see that the group with a low consumption of alcohol has higher grades on average than the group with high alcohol consumption:

```
In [19]:  fig, axes = plt.subplots(1,2, figsize = (14,4))
          sns.boxplot(x='acl', y='G3', data=student, ax=axes[0])
          sns.pointplot(x='acl', y='G3', data=student, ax=axes[1]);
```

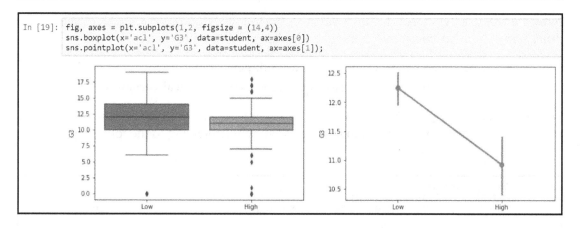

So let's set up the two hypotheses. For this question, our null hypothesis will be that for both groups, the population means of the final grades are equal, and the alternative hypothesis would be that the population means of the final grades are actually different. Before performing any tests, remember that all the tests have certain assumptions, and here are the three assumptions of the test that we are going to use to answer this question. You can verify that the three assumptions are fulfilled in this case, including the third one that variances are actually equal, and we just proved that with the test that we did previously.

In order to perform this t-test, we can use the `ttest_ind` function from the `stats` package. We pass the two vectors containing the grades for each group, and we give an additional argument that says that we actually have equal variances. When we perform this test, we get back the statistic and the *p*-value. We can see in the following screenshot that `pvalue` is very low; it is $4.6036088303692686e-06$ times 10 to -6, which is an extremely low number:

```
In [20]:  stats.ttest_ind(grades_low_acl, grades_high_acl, equal_var=True)

Out[20]:  Ttest_indResult(statistic=4.621320706949354, pvalue=4.603608830369269e-06)
```

In conclusion, this is very strong evidence in favor of the alternative hypothesis, which says that the population means in the grades are actually different when we compare these two groups. So this is our overall conclusion: there is a statistically significant difference between the grades of the two analyzed groups. Since the mean for the group with high alcohol consumption is less than the mean for the other group, the results suggest that alcohol consumption has a negative impact on the students' academic performance.

Now we will use hypothesis testing to answer the following question: *Do male teenagers drink more than female teenagers?*. Let's take a look at the following topics:

- Performing a **chi-squared** test for a contingency table
- Visualization of the relationship between two categorical variables

So, let's look at the Jupyter Notebook in the following screenshot to answer the question: do male teenagers drink more than female teenagers?, while referring to the following screenshot:

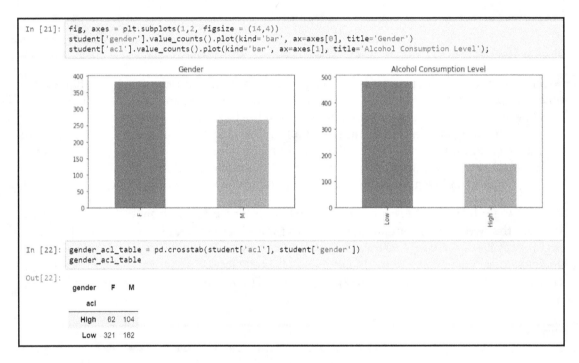

In the preceding screenshot, we see that we have more females than males, and we have more students with low alcohol consumption levels. There are many ways to perform, or to calculate, what is called a **contingency table** or a **cross-tabulation** table. We will use the `crosstab` function from `pandas` and we will pass the two series, and what we get back is the count for every category. When taking into account these two categorical variables, we also have the four possible categories. So we have 62 females with a high level of alcohol consumption, and 104 males for this group. We also have the count for the group of the low alcohol consumption females and males. Let's look at the following screenshot and visualize these numbers; we can visualize the proportions and we can also visualize the absolute counts:

```
In [23]:  fig, axes = plt.subplots(1,2, figsize = (14,4))
          gender_acl_table.plot(kind='bar', stacked=True, ax=axes[0]);
          (100*(gender_acl_table.T/gender_acl_table.apply(sum, axis=1)).T).plot(kind='bar', stacked=True, ax=axes[1]);
```

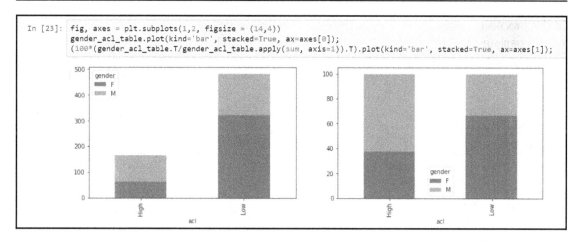

In the preceding screenshot, we can see that the visualization of the proportion of females in the low group is far greater than the proportion of males, and, on the other hand, for the high consumption group, the proportion of males is much greater. So it seems that we have a relationship between these two variables, and we definitely have a relationship in the sample. The question that we want to answer about the population is: *Can we say that this relationship holds for the population?* In order to answer this question, we have to perform a hypothesis test, as shown in the following screenshot:

```
In [24]:  chi_stat, p_value, dof, expected = stats.chi2_contingency(gender_acl_table)

In [25]:  p_value

Out[25]:  8.72933011769437e-11

In [26]:  expected_table = pd.DataFrame(expected, index=['High','Low'], columns=['F','M'])
          expected_table

Out[26]:
```

	F	M
High	97.96302	68.03698
Low	285.03698	197.96302

In the preceding screenshot, we see that the function that we will use in this case is from the `stats` package and is called `chi2_contingency` . We will pass the contingency table that we generated before. This function will give us back many objects, such as `chi-square statistic`, `p_value`, the degree of freedom, also known as `dof`, and a table with the expected values under the null hypothesis. In this case, the null hypothesis is that there is no relationship between these two variables in the population. Let's now run this test and look at the output shown in the following screenshot:

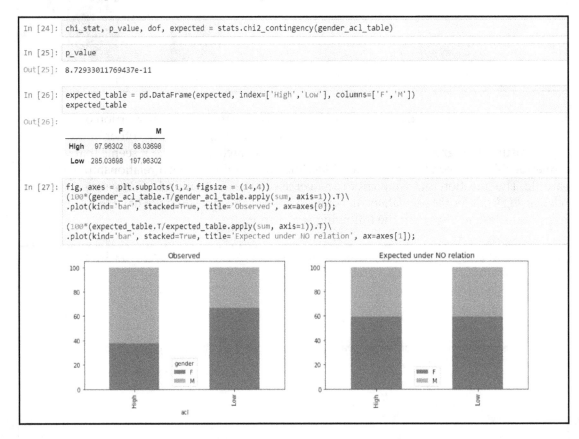

Let's take a look at the `p_value` object in the preceding screenshot. We can see that it is very low: `8.7293301176943706e-11` times 10 to -11. This is actually very good evidence in favor of the alternative hypothesis. In this case, the alternative hypothesis states that there is actually a relationship between these two variables. The other interesting object here is `expected`. This gives us the expected frequencies that we have observed if the null hypothesis was true. We must remember that the null hypothesis for this test is that there is no relationship between the two variables. So we can visualize in **In [32]** what our data would look like if there was no relationship in the population and make comparisons. Thus, this acts as additional evidence that shows us the difference between the drinking habits of male teenagers versus the drinking habits of female teenagers. We can see that male teenagers drink much more.

Summary

In this chapter, we started with an introduction to SciPy. We then had a brief overview of the various subpackages that are available to us. We learned how to perform a chi-square test for independence using the `scipy.stats` package. We used this package to perform other calculations, such as confidence intervals, probability calculations, and other types of statistical tests.

In the next chapter, we will introduce predictive analytics models, machine learning, the `scikit-learn` library, and more. Stay tuned!

6
Introduction to Predictive Analytics Models

In this chapter, we will take a closer look at the following topics:

- Predictive analytics
- Machine learning
- The scikit-learn library
- How to build a classification and a regression model

Predictive analytics and machine learning

In this section, we will learn the basics of predictive analytics. We will also take a closer look at the machine learning approach to predictive analytics. We will then proceed to discuss the various types of machine learning models. To end this section on a high note, we will look at the components of a supervised learning model.

 In this context, prediction does not necessarily refer to the future. It implies guessing something that is unknown or hasn't been observed yet.

There are many ways to predict something. We can begin by asking a high priest or a psychic, which is something humans have done for many millennia. Another approach could be to use your intuition or to ask an expert; these are more traditional business practices. There are various other ways to do this and one of the most recent and successful approaches is called **predictive analysis**.

The use of predictive analysis has increased drastically in recent years. There are two reasons for this:

- Technological advancements have given us the means to perform such analysis
- It is extremely effective, accurate, and works well

Predictive analytics is the use of data combined with techniques from mathematics, statistics, and computer science to make predictions about unknown events. The goal of predictive analytics is to produce a good assessment of what could happen regarding unknown events.

So, how do we carry out predictive analytics? There are many ways to perform predictive analytics, but one of the most successful tools for doing predictive analytics is machine learning. To understand how this works, we need to understand machine learning.

Machine learning is a sub-field of computer science. It can be simply described as giving a computer the ability to learn without it being explicitly programmed. The field of machine learning has developed many methods to teach computers to perform certain tasks using data. This approach has been very successful in doing predictive analytics. Machine learning also has a few drawbacks. It is essential to understand the problems one might face when using machine learning. We can separate these learning problems into a few broad categories. To simplify this, we shall consider two groups, called **supervised** and **unsupervised** learning. This section mainly focuses on supervised learning and the elements required to create a supervised learning model.

When thinking of creating a machine learning model, the first question that comes to mind is: *How does one know that they should use supervised learning*? The key to understanding this lies in understanding the target variable that needs to be predicted. Supervised learning has samples or observations about any particular concept/topic. Each observation will have different features, which are more commonly known as attributes or target variables. These target variables are what need to be predicted.

Let's consider an example of a dataset with a few rows of students, which that was used in `Chapter 5`, *Statistical Computing with Python*. The following table depicts gender, age, address, and various other characteristics:

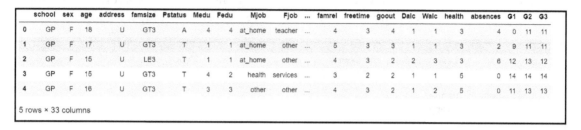

	school	sex	age	address	famsize	Pstatus	Medu	Fedu	Mjob	Fjob	...	famrel	freetime	goout	Dalc	Walc	health	absences	G1	G2	G3
0	GP	F	18	U	GT3	A	4	4	at_home	teacher	...	4	3	4	1	1	3	4	0	11	11
1	GP	F	17	U	GT3	T	1	1	at_home	other	...	5	3	3	1	1	3	2	9	11	11
2	GP	F	15	U	LE3	T	1	1	at_home	other	...	4	3	2	2	3	3	6	12	13	12
3	GP	F	15	U	GT3	T	4	2	health	services	...	3	2	2	1	1	5	0	14	14	14
4	GP	F	16	U	GT3	T	3	3	other	other	...	4	3	2	1	2	5	0	11	13	13

5 rows × 33 columns

If we define one of these features as the target variable that needs to be predicted, we would then use supervised learning.

Supervised learning has two major concerns: regression and classification. The difference between the two is simple. Classification occurs when the target variable is categorical in nature. Some examples of this would include alcohol consumption (low or high) or types of credit card transactions. On the other hand, if the target variable is a numerical variable, then we are talking about a regression problem. Examples of a numerical variable include the price of a house or stock, or the number of units sold in a month. For each of these classes of problems, there are many models that can be used to rectify them.

We can take away two points from this section. Firstly, we will treat the models in this section as a black box. This implies that I'm not going to explain any details about how they work internally. Considering that this is a high-level introduction to predictive analytics, we will look at the big picture on how to build predictive models. The same applies to the vast concept of machine learning.

Understanding the scikit-learn library

In this section, we will look at the `scikit-learn` library and will use it to implement a simple predictive model. To do this, we need to understand `scikit-learn` and how to load the `iris` dataset to the Jupyter Notebook. We will then take a closer look at how to build a supervised machine learning model using `scikit-learn` and, using this, we will build a simple predictive model.

scikit-learn

`scikit-learn` is the most popular Python library for doing machine learning. It provides a simple and efficient API with tools for data modeling and data analysis. It is built on top of NumPy, SciPy, and Matplotlib. The following is a screenshot of a Jupyter Notebook:

```
In [1]: from sklearn import datasets

In [2]: import pandas as pd
        import numpy as np

In [3]: iris = datasets.load_iris()
        iris_features = iris.data
        iris_target = iris.target

In [4]: iris_df = pd.DataFrame(data=iris.data, columns=iris.feature_names)
        iris_df['target'] = iris.target_names[iris.target]
        iris_df.head()
```

Out[4]:

	sepal length (cm)	sepal width (cm)	petal length (cm)	petal width (cm)	target
0	5.1	3.5	1.4	0.2	setosa
1	4.9	3.0	1.4	0.2	setosa
2	4.7	3.2	1.3	0.2	setosa
3	4.6	3.1	1.5	0.2	setosa
4	5.0	3.6	1.4	0.2	setosa

We do not import the entire library, but instead we import the ones we really need. We need to import the `datasets` objects. This allows us to load all the datasets that `scikit-learn` provides.

To understand the concept better, we will use the example of the `iris` dataset. This runs parallel to the *hello world* example for any machine learning datasets. The following table depicts a few rows from the `iris` dataset:

```
In [4]:  iris_df = pd.DataFrame(data=iris.data, columns=iris.feature_names)
         iris_df['target'] = iris.target_names[iris.target]
         iris_df.head()

Out[4]:
```

	sepal length (cm)	sepal width (cm)	petal length (cm)	petal width (cm)	target
0	5.1	3.5	1.4	0.2	setosa
1	4.9	3.0	1.4	0.2	setosa
2	4.7	3.2	1.3	0.2	setosa
3	4.6	3.1	1.5	0.2	setosa
4	5.0	3.6	1.4	0.2	setosa

This dataset contains a set of 150 observations of flowers; each flower has four measurements, which are the `sepal length`, `sepal width`, `petal length`, and `petal width`. In addition to this, we have the species of flower, which that we will use as the target variable.

The species mentioned throughout the dataset are `setosa`, `versicolor`, and `virginica`. If we use the measurements in the dataset to predict the species of the flower, which is a categorical variable, we are performing a classification task. The main API that is implemented in `scikit-learn` is that of the estimator. An estimator object is an object that contains the model that we can use to learn from the data.

The first thing to do in the Notebook is to import the estimator, more commonly known as the model. The following code is used to import the classifier:

```
from sklearn.neighbors import KNeighborsClassifier
```

The model that is used here is known as the KNeighbors model, which is imported from `scikit-learn`. We then proceed to create an instance of this object and it is called the `flower_classifier`. This is done by using the following line of code:

```
flower_classifier = KNeighborsClassifier(n_neighbors=3)
```

Here, we provide hyperparameters to the object. As mentioned before, these objects are being treated as black boxes and, hence, we will not go into depth as to why we are using a particular number or variable.

After this, we use the data that we have to train the estimator. To do this, we use the `fit` method of the `flower_classifier` object. This helps us pass the features and the target. This implies that we have used features to identify the target. The following code and its output are used to train the estimator:

```
In [10]:  flower_classifier.fit(X=iris_features, y=iris_target)

Out[10]:  KNeighborsClassifier(algorithm='auto', leaf_size=30, metric='minkowski',
                    metric_params=None, n_jobs=1, n_neighbors=3, p=2,
                    weights='uniform')
```

The model is now ready for evaluation. This is something we won't be taking a look at. Assuming the evaluation results are satisfactory, we can use this model to make `predictions`. The features used to predict the species of flowers are as follows:

```
In [11]:  # The features must be two-dimensional array
          new_flower1 = np.array([[5.1, 3.0, 1.1, 0.5]])
          new_flower2 = np.array([[6.0, 2.9, 4.5, 1.1]])


          0 == > setosa

          1 == > versicolor

          2 == > virginica

In [12]:  flower_classifier.predict(new_flower1)

Out[12]:  array([0])

In [13]:  flower_classifier.predict(new_flower2)

Out[13]:  array([1])

In [14]:  new_flowers = np.array([[5.1, 3.0, 1.1, 0.5],[6.0, 2.9, 4.5, 1.1]])
          predictions = flower_classifier.predict(new_flowers)
          predictions

Out[14]:  array([0, 1])
```

The arrays used must be two-dimensional NumPy arrays.

The output of the array when it is 0 classifies the species as a setosa. The numbers 1 and 2 classify the species as versicolor and virginica, respectively.

To predict the species of the flower, we use the classifier object and then pass new_flower1. The output of this is as follows:

```
In [12]:  flower_classifier.predict(new_flower1)

Out[12]:  array([0])
```

In accordance with the output received, the species that is labeled 0 is a setosa. In a similar fashion, we can predict the species of the second flower. We can continue to do the same for various measurements.

We can see a cumulative prediction by creating n two-dimensional NumPy arrays that contain the values of various flowers. We can call this the new_flowers function. A function named predictions is created, as follows:

```
In [14]:  new_flowers = np.array([[5.1, 3.0, 1.1, 0.5],[6.0, 2.9, 4.5, 1.1]])
          predictions = flower_classifier.predict(new_flowers)
          predictions

Out[14]:  array([0, 1])
```

As observed, the first value corresponds to the first flower and classifies it as a setosa, and the second value that corresponds to the second flower classifies it as a versicolor.

Building a regression model using scikit-learn

The previous section showed us an example of a classification model using `scikit-learn`. In this section, we will train a random forest model and use it to make predictions. We will also be building a classification model as the target variable in this scenario. This will be a categorical value that depicts the drinking habits of teenagers.

To do this, we first load the students dataset from the previous chapter. We will then train a logistic regression model and take a look at how to evaluate the classification model at a very basic level.

To begin, we load the libraries and import the students dataset and make some transformations to it, just as we did in the previous section. Our goal is to use the features of the students to predict the level of alcohol. These features are categorical values that can either be high or low. The code for loading the libraries and the dataset is as follows:

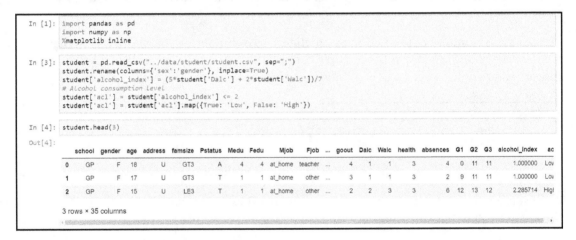

```
In [1]:  import pandas as pd
         import numpy as np
         %matplotlib inline

In [3]:  student = pd.read_csv("../data/student/student.csv", sep=";")
         student.rename(columns={'sex':'gender'}, inplace=True)
         student['alcohol_index'] = (5*student['Dalc'] + 2*student['Walc'])/7
         # Alcohol consumption level
         student['acl'] = student['alcohol_index'] <= 2
         student['acl'] = student['acl'].map({True: 'Low', False: 'High'})

In [4]:  student.head(3)
```

Out[4]:

	school	gender	age	address	famsize	Pstatus	Medu	Fedu	Mjob	Fjob	...	goout	Dalc	Walc	health	absences	G1	G2	G3	alcohol_index	ac
0	GP	F	18	U	GT3	A	4	4	at_home	teacher	...	4	1	1	3	4	0	11	11	1.000000	Lov
1	GP	F	17	U	GT3	T	1	1	at_home	other	...	3	1	1	3	2	9	11	11	1.000000	Lov
2	GP	F	15	U	LE3	T	1	1	at_home	other	...	2	2	3	3	6	12	13	12	2.285714	Higl

3 rows × 35 columns

The following is a list of all the features we would like to use for our classification model:

```
In [5]:  features = ['gender','famsize','age','studytime','famrel','goout','freetime','G3']
         target = 'acl'
```

The `scikit-learn` library only understands numbers. This makes it important to convert variables that are not numeric in nature into numeric values. To do this, you use dummy features that are more commonly known as one-hot encoding.

To one-hot encode our variables, the females will correspond to 0 and the males will correspond to 1. We will perform the same transformation for the family size feature, and for the alcohol consumption level; the low will get assigned to 0 and the high will get assigned to 1. The code should look something like the following:

```
In [6]:  # For gender: Female will be 0, Male will be 1
         student['gender'] = student['gender'].map({'F':0, 'M':1}).astype(int)
         # For famsize: 'LE3' - less or equal to 3 will be 0. 'GT3' - greater than 3 will be one
         student['famsize'] = student['famsize'].map({'LE3':0, 'GT3':1}).astype(int)
         # for acl: 'Low' will be 0, 'High' will be 1
         student['acl'] = student['acl'].map({'Low':0, 'High':1}).astype(int)
```

B11522_6_013

We then save these values into objects X and Y by using the following code:

```
X = student[features].values
y = student[target].values
```

To build a simple model, we need to predict the most common category. In this example, we can find the most common category by using the following code:

```
In [8]:  student['acl'].value_counts(normalize=True)

Out[8]:  0    0.744222
         1    0.255778
         Name: acl, dtype: float64
```

This clearly implies that about 74% of students reported a low level of alcohol consumption. Thus, we can build a simple model that correctly classifies 74% of unseen cases. This number is important because it gives us the first benchmark that we can use to compare how good the models that we build are.

We now proceed to build a predictive model called logistic regression. We will import the model, create an instance of this object, and then train the model with our data. The following code block depicts how this is done:

```
In [9]:   from sklearn.linear_model import LogisticRegression

In [10]:  student_classifier_logreg = LogisticRegression(C=2)

In [11]:  student_classifier_logreg.fit(X, y)
Out[11]:  LogisticRegression(C=2, class_weight=None, dual=False, fit_intercept=True,
                   intercept_scaling=1, max_iter=100, multi_class='ovr', n_jobs=1,
                   penalty='l2', random_state=None, solver='liblinear', tol=0.0001,
                   verbose=0, warm_start=False)
```

To evaluate this model thoroughly, we need to evaluate the model with cross-validation. The evaluation that we will perform in this section is naive or basic in nature. It is done so as to give you an overview or a general idea of the concept. The general idea is the following:

```
In [12]:  student['predictions_logreg'] = student_classifier_logreg.predict(X)

In [13]:  confusion_matrix = pd.crosstab(student['predictions_logreg'], student['acl'])
          confusion_matrix
Out[13]:
```

acl	0	1
predictions_logreg		
0	453	105
1	30	61

We begin by calculating the predictions in the cell that calculates the predictions the model makes. We then tabulate these predictions with the actual observations and build something that is called a confusion matrix:

acl	0	1
predictions_logreg		
0	453	105
1	30	61

The diagonal of the matrix depicts the number of cases where the classifier made a correct prediction. We can then use these numbers to calculate a simple evaluation matrix called `Accuracy`. This is simply the fraction of correct predictions that were produced by the model. So, out of all of these predictions, the values at the top left and bottom right are the ones that are correct.

Using this, we can find the accuracy of the model. The following code helps us to do this:

```
In [14]:  ac = (confusion_matrix.ix[0,0] + confusion_matrix.ix[1,1])/student.shape[0]
          print("Accuracy: {}".format(ac))

          Accuracy: 0.7919876733436055
```

It is observed that the accuracy of our model is 0.79 or 79%. We compare this value with the accuracy of our simple model, which was 74%. The values obtained from the two models do not vary by much. The reason for this could be that the logistic regression model is a rather simple model.

We can try using a more complex model called `RandomForestClassifier`. We shall use this model as a black box just to see the variation in accuracy. We begin by importing an object, creating an instance of that object, training the model with the data, and then producing predictions. The following code block is used to do this:

```
In [14]:  from sklearn.ensemble import RandomForestClassifier

In [15]:  student_classifier_rf = RandomForestClassifier()

In [16]:  student_classifier_rf.fit(X,y)
          student['predictions_rf'] = student_classifier_rf.predict(X)

In [17]:  confusion_matrix = pd.crosstab(student['predictions_rf'], student['acl'])
          confusion_matrix

Out[17]:
              acl    0    1

          predictions_rf

                0   480   19

                1    3   147

In [18]:  ac = (confusion_matrix.ix[0,0] + confusion_matrix.ix[1,1])/student.shape[0]
          print("Accuracy: {}".format(ac))

          Accuracy: 0.9661016949152542

In [19]:  # ['gender', 'famsize', 'age', 'studytime', 'famrel', 'goout', 'freetime', 'G3']
          new_student = np.array([[0, 1, 18, 2, 1, 5, 5, 16]])
          prediction = student_classifier_rf.predict(new_student)
          print("The model predicts that the student belongs to the:")
          if prediction == 1:
              print("HIGH Alcohol Consumption group")
          else:
              print("LOW Alcohol Consumption group")
```

Here, we build a new confusion matrix, as follows, and it is observed that the accuracy of this model is high:

acl	0	1
predictions_rf		
0	480	19
1	3	147

At the end of our code, we achieve an accuracy of 96%. The reason for this is that complex models such as random forest often carry out overfitting. Overfitting is the concept where the model learns what is going on in the dataset, but this knowledge doesn't generalize unseen data.

This is one of the major reasons why model evaluation is very complex in nature and requires cross-validation. However, the accuracy matrix is one that is often used with predictive models.

We can now proceed to use this model to make predictions about unseen data. Assume we have a new student with various features. In this instance, the student is a male with a large family and is 18 years of age. He studies for 2 hours per week, his family relations are not very good, and so on. We can use the following code to find out the level of alcohol he consumes:

```
In [20]: # ['gender', 'famsize', 'age', 'studytime', 'famrel', 'goout', 'freetime', 'G3']
         new_student = np.array([[1, 1, 18, 2, 1, 5, 5, 10]])
         prediction = student_classifier_rf.predict(new_student)
         print("The model predicts that the student belongs to the:")
         if prediction == 1:
             print("HIGH Alcohol Consumption group")
         else:
             print("LOW Alcohol Consumption group")

         The model predicts that the student belongs to the:
         HIGH Alcohol Consumption group
```

The model then predicts that the student belongs to the `HIGH Alcohol Consumption group`.

Let's assume another case study, where the student is a female, she does very well in school and has a final grade of 16. The following screenshot depicts the output of the code:

```
In [19]: # ['gender', 'famsize', 'age', 'studytime', 'famrel', 'goout', 'freetime', 'G3']
         new_student = np.array([[0, 1, 18, 2, 1, 5, 5, 16]])
         prediction = student_classifier_rf.predict(new_student)
         print("The model predicts that the student belongs to the:")
         if prediction == 1:
             print("HIGH Alcohol Consumption group")
         else:
             print("LOW Alcohol Consumption group")

         The model predicts that the student belongs to the:
         LOW Alcohol Consumption group
```

The model predicts that a student with these characteristics will belong to the `LOW Alcohol Consumption` group.

Regression model to predict house prices

In this section, we will build a regression model using the housing dataset from the previous sections. We begin by loading the housing prices dataset and preparing it for modeling. We then train a linear regression model and proceed to evaluate this model in a simple but intuitive manner. We shall conclude by using this model to make predictions.

We load the libraries that we will need to use and also import the dataset. As observed in previous sections, we are aware of the fact that there are a number of neighborhoods in this dataset that contain very few observations. To eliminate this, we would use this model only for neighborhoods with more than 30 observations. To do this, we need to use the following code block:

```
counts = housing['Neighborhood'].value_counts()
more_than_30 = list(counts[counts>30].index)
housing = housing.loc[housing['Neighborhood'].isin(more_than_30)]
```

Based upon the exploratory data analysis from previous sections, we will choose the following features for our model:

```
features = ['CentralAir', 'LotArea', 'OverallQual', 'OverallCond',
 '1stFlrSF', '2ndFlrSF', 'BedroomAbvGr', 'Age']
target = 'SalePrice'
```

It is observed that the target variable will be the SalePrice of a house, and as this is a numerical variable, we are dealing with a regression problem.

It is also common knowledge that the two features, Neighborhood and CentralAir, are not numeric in nature, thus, they need to be transformed into numerical variables. To do this, we use the following lines of code:

```
# Neighborhood
dummies_nb = pd.get_dummies(housing['Neighborhood'], drop_first=True)
housing = pd.concat([housing, dummies_nb], axis=1)
# CentralAir
housing['CentralAir'] = housing['CentralAir'].map({'N':0,
'Y':1}).astype(int)
```

These two lines of code create a new variable, called a dummy variable, for each neighborhood. This essentially creates a new vector for each neighborhood that has a value of 1. This is the case if the house belongs to that neighborhood. If the house does not belong to that neighborhood, it creates a new vector with a value of 0. We also assign 0 to the houses that don't have CentralAir, and 1 to the houses that have CentralAir.

We then add these new features to our list of features and create objects that contain our features, target variable, and the number of observations:

```
In [6]:  features += list(dummies_nb.columns)
```

```
In [7]:  X = housing[features].values
         y = housing[target].values
         n = housing.shape[0]
```

Before we train our model, we need to make sure it's simple. For a regression model, the simplest possible model just predicts the average. To achieve the average value we use the following code:

```
In [8]:  y_mean = np.mean(y)
         y_mean
```

```
Out[8]:  180167.63358778626
```

The average for our target variable is an approximate of $180,000. For the model evaluation, we will use the evaluation metric called the root-mean-square error, commonly known as RMSE. The formula is as follows:

$$RMSE = \sqrt{\frac{\sum(obs - pred)^2}{n}}$$

It compares the observed values with the predicted values. The closer the observed and the predicted values are, the smaller the metric. We want the smallest metric possible.

To calculate the root-mean-square error for our simple model, we will use the following lines of code:

```
In [9]:  RMSE_null_model = np.sqrt(np.sum((y - y_mean)**2) / n)
         RMSE_null_model

Out[9]:  78032.944854541085
```

We observe that this value is approximate of $70,000.

To build a regression model, we first import an object and then create an instance of that object. We train the model and make predictions:

```
In [10]:  from sklearn.linear_model import LinearRegression

In [11]:  regressor = LinearRegression()

In [12]:  regressor.fit(X, y)

Out[12]:  LinearRegression(copy_X=True, fit_intercept=True, n_jobs=1, normalize=False)

In [13]:  housing['predictions'] = regressor.predict(X)

In [14]:  y_pred = housing['predictions'].values
```

To calculate the root-mean-square error of the model, we need to implement the following lines of code:

```
In [15]:  RMSE_regressor = np.sqrt(np.sum((y - y_pred)**2) / n)
          RMSE_regressor

Out[15]:  33729.218173366113
```

The output of this model is around $33,000, which is a significantly lower number in comparison to our simple model. To visualize and compare the `predictions` that we have with the real `SalePrice` of the houses, we can create a scatter plot by using the following function:

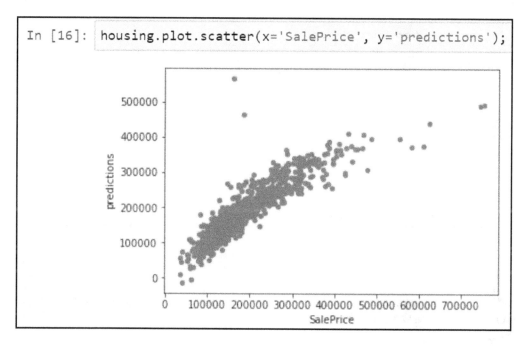

```
In [16]:  housing.plot.scatter(x='SalePrice', y='predictions');
```

The scatter plot clearly depicts that the predictions are very close to the actual `SalePrices` of the houses. This proves that the model is accurate.

Moving on, we will predict the value of a new house with various characteristics. The house has `CentralAir`, `LotArea`, `OverallQual` of 6, and `OverallCond` of 6. The code for predicting the value of this house is as follows:

```
In [17]:  new_house = np.array([[0, 12000, 6, 6, 1200, 500, 3, 5, 0,0,1,0,0,0,0,0,0,0,0,0,0,0]])
          prediction = regressor.predict(new_house)
          print("For a house with the following characteristics:\n")
          for feature, feature_value in zip(features, new_house[0]):
              if feature_value > 0:
                  print("{}: {}".format(feature, feature_value))
          print("\nThe predictied value for the house is: {:,}".format(round(prediction[0])))

          For a house with the following characteristics:

          LotArea: 12000
          OverallQual: 6
          OverallCond: 6
          1stFlrSF: 1200
          2ndFlrSF: 500
          BedroomAbvGr: 3
          Age: 5
          Edwards: 1

          The predictied value for the house is: 184,395.0
```

The model predicts that this is from the `Edwards` neighborhood and has a value of `188,455`.

For a different example, you can change the neighborhood, assuming that this house doesn't have `CentralAir` and belongs to the `Timber` neighborhood. The output of the code will be as follows:

```
In [18]:  new_house = np.array([[0, 12000, 6, 6, 1200, 500, 3, 5, 0,0,0,0,0,0,0,0,0,0,0,0,0,1]])
          prediction = regressor.predict(new_house)
          print("For a house with the following characteristics:\n")
          for feature, feature_value in zip(features, new_house[0]):
              if feature_value > 0:
                  print("{}: {}".format(feature, feature_value))
          print("\nThe predictied value for the house is: {:,}".format(round(prediction[0])))

          For a house with the following characteristics:

          LotArea: 12000
          OverallQual: 6
          OverallCond: 6
          1stFlrSF: 1200
          2ndFlrSF: 500
          BedroomAbvGr: 3
          Age: 5
          Timber: 1

          The predictied value for the house is: 214,944.0
```

The model then predicts that the value of the house is £214,944. We can use the same code to produce predictions for different houses. Now, please keep in mind that the models that we built in this section were not thoroughly evaluated. Although we have used the popular metrics for evaluation, a key step in machine learning modeling is to use cross-validation. This is absolutely essential for a good assessment of the performance of models.

Summary

In this chapter, we learned about predictive analytics and some concepts related to supervised machine learning, and gained essential knowledge about how to perform machine learning with Python. We also took a look at various practical applications of predictive analytics using `scikit-learn`. We took a closer look at how to train a classification model and then we used it to make predictions. We also built a regression model and used it to make predictions.

Other Books You May Enjoy

If you enjoyed this book, you may be interested in these other books by Packt:

Hands-On Data Analysis with NumPy and Pandas
Curtis Miller

ISBN: 978-1-78953-079-7

- Understand how to install and manage Anaconda
- Read, sort, and map data using NumPy and pandas
- Find out how to create and slice data arrays using NumPy
- Discover how to subset your DataFrames using pandas
- Handle missing data in a pandas DataFrame
- Explore hierarchical indexing and plotting with pandas

Beginning Data Science with Python and Jupyter
Alex Galea

ISBN: 978-1-78953-202-9

- Identify potential areas of investigation and perform exploratory data analysis
- Plan a machine learning classification strategy and train classification models
- Use validation curves and dimensionality reduction to tune and enhance your models
- Scrape tabular data from web pages and transform it into Pandas DataFrames
- Create interactive, web-friendly visualizations to clearly communicate your findings

Leave a review - let other readers know what you think

Please share your thoughts on this book with others by leaving a review on the site that you bought it from. If you purchased the book from Amazon, please leave us an honest review on this book's Amazon page. This is vital so that other potential readers can see and use your unbiased opinion to make purchasing decisions, we can understand what our customers think about our products, and our authors can see your feedback on the title that they have worked with Packt to create. It will only take a few minutes of your time, but is valuable to other potential customers, our authors, and Packt. Thank you!

Index